THE EMERGING TECHNOLOGY OF BIG DATA

Its Impact as a Tool of ICT Development

THE EMERGING TECHNOLOGY OF BIG DATA

Its Impact as a Tool of ICT Development

Heru Susanto, PhD
Fang-Yie Leu, PhD
Chin Kang Chen, PhD

AAP APPLE ACADEMIC PRESS

Apple Academic Press Inc.
3333 Mistwell Crescent
Oakville, ON L6L 0A2
Canada

Apple Academic Press Inc.
1265 Goldenrod Circle NE
Palm Bay, Florida 32905
USA

Exclusive worldwide distribution by CRC Press, a member of Taylor & Francis Group

International Standard Book Number-13: 978-1-77188-675-8 (Hardcover)
International Standard Book Number-13: 978-1-351-24125-0 (eBook)

Library and Archives Canada Cataloguing in Publication

Title: The emerging technology of big data : its impact as a tool of ICT development / Heru Susanto, PhD, Fang-Yie Leu, PhD, Chin Kang Chen, PhD.

Names: Susanto, Heru, 1977- author. | Fang-Yie, Leu, author. | Chen, Chin Kang, author.

Description: Includes bibliographical references and index.

Identifiers: Canadiana (print) 20189068469 | Canadiana (ebook) 20189068477 | ISBN 9781771886758 (hardcover) | ISBN 9781351241250 (PDF)

Subjects: LCSH: Big data. | LCSH: Information technology. | LCSH: Communication and technology. | LCSH: Data mining. | LCSH: Electronic data processing—Data preparation. | LCSH: Relationship marketing.

Classification: LCC QA76.9.B45 S87 2018 | DDC 005.7—dc23

Library of Congress Cataloging-in-Publication Data

Names: Susanto, Heru, 1965- author. | Fang-Yie, Leu, author. | Chen, Chin Kang, author.

Title: The emerging technology of big data : its impact as a tool of ICT development / Heru Susanto, PhD, Fang-Yie Leu, PhD, Chin Kang Chen, PhD.

Description: Toronto ; New Jersey : Apple Academic Press, 2019. | Includes bibliographical references and index.

Identifiers: LCCN 2018058124 (print) | LCCN 2019000439 (ebook) | ISBN 9781351241250 (ebook) | ISBN 9781771886758 (hardcover : alk. paper)

Subjects: LCSH: Big data. | Information technology.

Classification: LCC QA76.9.B45 (ebook) | LCC QA76.9.B45 S87 2019 (print) | DDC 005.7--dc23

LC record available at https://lccn.loc.gov/2018058124

ABOUT THE AUTHORS

Heru Susanto, PhD

Heru Susanto, PhD, is currently Head of the Information Department at the Indonesian Institute of Sciences, Computational Science & IT Governance Research Group, where he is also a researcher. He is also an Honorary Professor and Visiting Scholar at the Department of Information Management, College of Management, Tunghai University, Taichung, Taiwan. Dr. Susanto has worked as an IT professional in several roles, including Web Division Head of IT Strategic Management at Indomobil Group Corporation and Prince Muqrin Chair for Information Security Technologies at King Saud University. His research interests are in the areas of information security, IT governance, computational sciences, business process re-engineering, and e-marketing. Dr. Susanto received a BSc in Computer Science from Bogor Agricultural University, an MBA in Marketing Management from the School of Business and Management Indonesia, an MSc in Information Systems from King Saud University, and a PhD in Information Security System from the University of Brunei and King Saud University.

Fang-Yie Leu, PhD

Fang-Yie Leu, PhD, is currently a Professor in the Computer Science Department, Tunghai University, Taiwan, and on the editorial boards of several journals. Prof. Leu organizes Mobile Commerce, Cloud Computing, Network and Communication Security (MCNCS) and Compact Wireless EVA Communications System (CWECS) international workshops. He is an IEEE member and now serves as the Technical Program Committee (TPC) member of many international conferences. He was also a visiting scholar of Pittsburgh University, Pennsylvania. His research interests include wireless communication, network security, grid applications and security. Dr. Leu received his bachelor, master, and PhD degrees all from the National Taiwan University of Science and Technology, Taiwan.

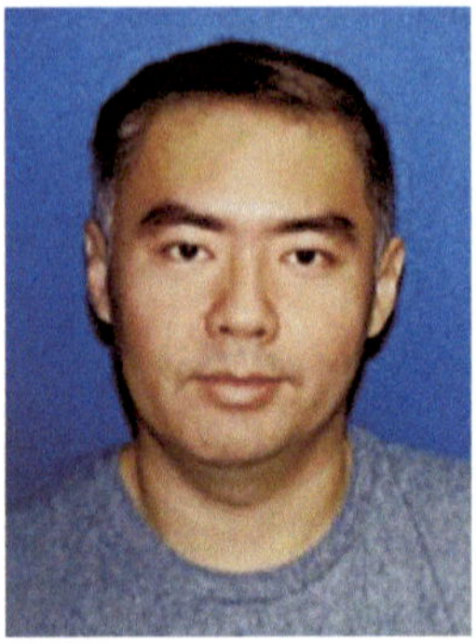

Chin Kang Chen, PhD

Chin Kang Chen, PhD, is a Lecturer at the Universiti Brunei Darussalam, Begawan, Brunei. He also has industry experience from working on projects with the local telecommunications companies and government ministries and agencies. His research interests include information security management system, IT governance, mobile devices, cloud computing and e-government. Dr. Chen received his bachelor of Engineering from the Royal Melbourne Institute of Technology, Australia; his Master of Digital Communication from Monash University; and his Master of Information Technology from Queensland University of Technology; and his PhD from the University of Brunei.

CONTENTS

ABBREVIATIONS

AI	artificial intelligence
API	application programming interface
AWS	Amazon Web store
BI	business intelligence
BPR	business process reengineering
CCTV	closed circuit television
CMDB	configuration management databases
CPS	cyber-physical system
CRM	customer relationship management
DBMS	database management system software
DIAD	delivery information acquisition device
DRP	distribution resource planning
EDI	electronic data interchange
ERP	enterprise resource planning
Gen Y	generation Y
Gen Z	generation Z
GIS	geographic information system
GISMO	graphical interactive student monitoring tool
GPA	grade point average
HR	human resource management
ICT	information and communication technology
iNEIS	Integrated National Education Information System
IoT	internet of things
IS	information system
IT	information technology
LMS	learning management system
ME	mechanical engineering
MIS	management information systems
ML	machine learning
MPT	microblog processing toolkit
OLPIT	ontology for linking processes and IT infrastructure
ORION	On-Road Integrated Optimization and Navigation
POS	point-of-sale

QMP	quality management practices
RFID	radio-frequency identification
ROI	return of investment
SaaS	software as a service
SIS	student information system
SQL	structured query language
UCAS	Universities and Colleges Admission Service
UNESCO	United Nations Educational, Scientific, and Cultural Organization
UNWE	University of National and World Economy
UPS	United Parcel Service
VCA	video content analysis
VLE	virtual learning environment
VPE	virtual production environment

PREFACE

In today's modern society, information system (IS) through big data contributes to the success of organizations, as it gives a solid foundation to increase both efficiency and productivity. Many business organizations realize that compliance with big data regulations will affect their business prospects. An IS is a group of computer tools that can be used for collecting, storing, or processing data. Institutions depend on these systems to help enable and support their operations, interact with people, manage their workforce, provide services, and many other processes. Every day, the amount of data collected in the digital tools grows more tremendously than in previous years. As the amount of data increases, the use of IS becomes more essential, and thus, with every passing year, the price of IS hardware has decreased significantly as well. This development is occurring under the Moore's law, as the heart of microprocessors has been doubling every 18–24 months.

To manage these data, we depend on IS to create, control, store, distribute, locate, and access this information. However, traditional computing solutions are not scalable enough to manage such magnitude of data. These large sets of data volume are known as big data. Big data is the result of collecting a large volume of data across many sites. It is important to have the right information system to cater to the high volume of data collected in a system. Academic institutions can benefit from big data if they know how to manage it accurately.

Big data is often determined according to the "3Vs," which is volume, variety, and velocity. The 3Vs can also be described as the quantity, diverse types of data, and rate of flow of information going into organizations that exceeds the capacity of a traditional computer. The volume aspect of 3Vs is the size of data. It is difficult to determine the limits of big data, so this aspect is very relative in the education field. Variety is different types or formats of big data. Therefore, this means that big data in academic settings collects, analyzes, and provides information with different backgrounds to ensure better learning resources for institutions. Last, velocity is the increasing flow of data and the need for hardware in ISs to carry more and more information, and for software to process these data as quickly as possible. Big data ensure that stakeholders in the academic environment can have a quick access to information needed in their educational processes.

There are many challenges faced in handling big data. For many years, the collection of big data is still growing massively, and until now, leaders are still figuring out how to deal with these particular challenges. The challenges do not only focus mainly on the volume of data, but also from variety and velocity. Thus, to manage big data effectively, leaders must also assess these issues. In the case of an academic institution, data volume problems such as an overbearing amount of data can come from accumulation of old types of data that have been stored for many years plus the additional new types of data.

Shrinking size of data can be exampled by Twitter interactions. Most of the interactions present on Twitter are mainly texts, which can be compressed easily at high rates. In terms of use for academic purposes, many data available could also be modified in the same way as Twitter, with the fact that they are mainly consisting of texts.

This study highlights, emphasizes, and analyzes the impacts of big data and how they can be implemented in the areas of social media, business process re-engineering, science, e-learning, higher education, business intelligence, and green information and communication technologies. The result is very promising as big data is now highly regarded and accepted as a useful tool to help organizations systematically to manage their data and information effectively and efficiently.

CHAPTER 1

MANAGING BIG DATA'S IMPACT ON SOCIAL MEDIA FOR ORGANIZATION BUSINESS PROCESS REENGINEERING

HERU SUSANTO

The Indonesian Institute of Sciences, Indonesia,
E-mail: heru.susanto@lipi.go.id

Department of Information Management, College of Management, Tunghai University, Taichung, Taiwan

ABSTRACT

Social media has been become one of the largest platforms on the Internet for people to interact with each other. It can be used as a platform to further connect with consumers. Social media means any form of websites or applications that allows users to create and share information actively or to participate in social networking. People, including students, use social media in all sorts of different ways. One of the growing trends in social media is to start a business online, a form of e-commerce, whereby the social media can serve as a platform or an online retail store for certain products. With the emergence and evolution of the smartphones that are equally competitively produced by Apple, Samsung, mobile phones have become one of the most globally popular social media platforms that provide various service of communication. The power of social networking is very interesting, as it not only serves as a platform for networking with others, but it can also be used as a hub for educational learning. In business practices, companies are able to make use of social media to reach out and connect with their customers in order to ensure consumers of their products maintain a sense of loyalty. Social media consists of unlimited data, called by big data. Big data is defined as a cultural, technological, and scholarly phenomenon that rests on the interplay of technology and analysis and mythology. These three concepts are linked together when technology maximizes the computation power and algorithmic accuracy to gather, analyze, link and compare large data sets. Analysis works in a way that it draws large data sets to identify the patterns in order to make economic, social, technical, and legal claim. It is a widespread myth that large data sets offer a higher form of intelligence and knowledge that can generate insights that were previously impossible, with the aura of truth, objectivity, and accuracy. This study emphasized the effect of big data and social media issues for organization redesign of business processes and reengineering.

1.1 INTRODUCTION

Today, social media has been the driving force behind modernization. It has become one of the largest platforms on the Internet for people to interact with each other. In terms of business perspective, it can be used as a platform to further connect with consumers. Big data, on the other hand, is the amount of data that is readily available over the Internet; it consists of everything that an individual inputs onto the Internet. Taking social media as an example, on Facebook when an individual updates his/her status, it will be saved onto the server as part of the database. This is true for Twitter as well, when one is to update their status; the information is uploaded onto the server and stored.

Kirschner and Karpinski (2010) reported a negative relation between the usage of Facebook and a student's grade point average (GPA). Through the collection of quantitative data, the data showed that there were mean differences between the GPAs of active Facebook users, averaging 3.06 GPA, and non-active Facebook users, averaging 3.82 GPA. Thirty-five of the 219 participants provided a qualitative data by stating their reasons for the impact of Facebook on their performance. Those suggesting that Facebook had a negative impact had comments about distraction or poor time-management skills.

There are three ways in which privacy can be invaded online. The first was an uninvited intrusion into a user's personal space. This includes online marketing, spam advertising, pop-ups and sponsored sites around the edges of a web page. The study shows that such intrusions are the most noticeable invasion of privacy in people's minds, even though their potential consequences are the least harmful. The second threat which is believed to be the most serious is fraudulent e-commerce transactions and identity theft. Although it is the most serious, the case study shows that people are not worried about such activities as they believed that it would be done by big data companies such as Google and Facebook.

The third kind of privacy invasion is personal profiling by big companies for commercial advantages, whereby big data companies such as Google or Facebook combine hundreds and thousands of data obtained from different sources, known as the process of data blending, to understand who an individual is, where he/she lives, where they usually go, who their friends are, what they prefer to buy, etc.. This information may be simply used to make offers that are likely to appeal to us as individuals or for less harmful purposes such as knowing whether an individual engages in risky hobbies and should be charged with higher insurance rates.

This study discusses the effect of big data and social media issues for business processes and reengineering. The discussion on social media and big data is provided in Sections 1.2 and 1.3. The emerging technology issues and effect of big data and social media is discussed in Sections 1.4 and 1.5. Section 1.6 provides information on managing change through big data and social media. Finally, the conclusion is provided in Section 1.7.

1.2 SOCIAL MEDIA

Social media is defined as a group of Internet-based applications that allow the creation and exchange of user-generated content (Kaplan and Haenlein, 2010). Social media, in general, means any form of websites or applications that allows users to create and share information actively or to participate in social networking. People use social media in all sorts of different ways such as students, for example, make use of social media to learn. One of the growing trends in social media is to start a business online, a form of e-commerce, whereby the social media can serve as a platform or an online retail store for certain products.

From the definition, we can deduce that social media typically consists of tools such as blogs, social network sites, virtual worlds, games, Wikipedia, YouTube, etc. Social media is a category of which emerged online media incorporates with a number of characteristics such as participation, openness, conversation, community, and connectedness, all of which requires the abovementioned tools to execute.

It is worth noting that social media is not only found on the Internet. In recent years, with the emergence and evolution of the phones that is equally competitively produced by Apple, Samsung, and so forth, mobile phones have become one of the most globally popular social media platforms that provide various service of communication. An example of a popular form of communication on social media through the use of mobile phones is the application "WhatsApp" with an estimated 900 million users worldwide on a monthly basis, it has become arguably one of the most popular applications to date, and continues to offer its services to the world for free (Susanto, 2018; Almunawar et al., 2018a; Almunawar et al., 2018b; Susanto et al., 2016a; Almunawar et al., 2015).

A few more examples of social media would be, "Facebook," one of the first social networks was created by university dropout Mark Zuckerberg. Facebook has a whooping 1.55 billion monthly active users on the application as of January 2016, that accounts for roughly 21% of the world's population alone, with other applications such as Twitter and Instagram, having 320 million and 400 million users, respectively.

The power of social network is very interesting, as it not only serves as a platform for networking with others, it can also be used as a hub for educational learning. With the development and emergence of Web 2.0, which offers interactive media that is easily accessible and free, an increasing number of users have progressed it from a media for sharing to a tool for learning together. The usage of Google Drive is a good example of this, whereby an individual can create a document and then share it with others to work on the document together without having to meet each other. Social media has made it promising to improve learning and teaching experience among the students and teachers.

In business practices, companies are able to make use of social media to reach out and connect with their customers in order to ensure consumers of their products maintain a sense of loyalty. Brand loyalty is a signal that intensifies consumers' trust and the relationship that a consumer shares with the brand itself and how they identify themselves with the brand. As a brand creates an exclusive and premium product, it produces significant means to human mind in order to make their products irresistible and irreplaceable and eventually winning the loyalty of consumers. This, in return, is able to increase sales revenue, market share, profitability to firm and helps with the growth of firm as well as maintaining their position in the marketplace (Almunawar et al., 2013a; Almunawar et al., 2013b).

The study, based on brand loyalty, which focused on Turkey, found that the driving force behind the brand loyalty was indeed an advantageous campaign on social media and also the campaign appearing on different kinds of social media platforms.(Erdoğmuş and Cicek, 2012)

1.3 BIG DATA

Big data is defined as a term that describes a large and complex amount of data both structured and unstructured inundate in business on a daily basis. Big data is a relatively new data that has only come about in the 21st century whereby the concept of big data only gained momentum in the early 2000s when an industry analyst, Doug Laney, divided the concept distinctly into three parts which is known as the three Vs: volume, velocity, and variety. Volume is where an organization collects data from a variety of sources, including business transactions, social media, and information from sensor to machine-to-machine data. In the past, storing it would have been a problem, but the new technologies, such as Hadoop, have eased the burden. Velocity is where the data streams in at an unprecedented speed, and must be dealt with in a timely manner. RFID tags, sensors, and smart metering are driving the need to deal with torrents of data in near-real time. Variety is when data comes in all different kinds of formats ranging from structured, numeric data in traditional databases to unstructured text documents, e-mail, video, audio, stock ticker data, and financial transactions.

Big data can also be defined as a cultural, technological, and scholarly phenomenon that rests on the interplay of technology and analysis and mythology. These three concepts are linked together when technology maximizes the computation power and algorithmic accuracy to gather, analyze, link and compare large data sets. Analysis works in a way that it draws large data sets to identify the patterns in order to make economic, social, technical, and legal claim. Mythology is the widespread belief that large data sets offer a higher form of intelligence and knowledge that can generate insights that were previously impossible, with the aura of truth, objectivity, and accuracy (Susanto et al., 2018).

Big data can be divided into two types, the first being public data and the other private data. Public data is information that is readily available for access by the public. They can be freely used, reused, and redistributed by anyone with no existing local, national, or international legal restrictions on access or usage. Private data is basically the opposite of public data, where the data is not available to the public but is held by and available to the companies that have the data. However, these data can be transferred to third parties provided that users have given permission or that the data has been anonymized whereby the identifying information has been removed. In short, data for researches or commercials relies on how users have customized their contents to be shown to the public which can be seen by anonymous or private parties where the user identifies who can see it.

The term "big data" is being thrown around today and can generally be confusing for the most people. Big data has been used as a term to describe different kinds of ideas such as huge quantities of data, social media analytics, next generation data management capabilities, real-time data, and so forth. Whichever ways they are labeled, organizations have begun to slowly understand the capability of analyzing this vast array of data and as such, there has been a group of growing pioneers that is achieving a breakthrough success in business outcomes, especially those that make use of social media. Organizations have now understood that in order to compete economically worldwide, they need to comprehensively understand the markets, consumers, products, regulations, competitors, suppliers, employees, and so forth. All these factors require the effective usage of information and analytics and the reason why the information is considered a valuable asset for a company. With the emergence of big data, organizations are discovering new ways to be more competitive and to beat the market by improving themselves in all the aspects that are beneficial for their businesses. However, they are not all able to make full use of such information, but it is definitely available with the industry.

The benefits of big data are not to be underestimated. It can benefit major companies that can use such data to their advantages. A telecom chain is one such business that could benefit. With big data, they are able to route optimally and increase the quality of service by analyzing their network in real time. They can also cope with fraud by analyzing call data records in real time. It is also possible for them to be able to increase their profits as it allows the call centers representatives to be able to flexibly modify the subscriber calling plans immediately. Telecoms are also able to tailor their marketing campaigns to suit individual consumers based on the location and social network technologies. Last, they are able to use big data and create an insight into consumers' behavior and be able to use the data to develop new products and services to better improve their businesses (Acker *et al.*, 2013).

1.4 EMERGING AFFECTED

With huge user base on social media applications such as Facebook, it is no surprise that a huge amount of data is generated which means that inevitably social media and big data are linked directly with each other. Statistically, it has been shown that people from around the world post 400 million tweets on Twitter, adding to the sum of 350 million pictures on Facebook and 4 billion views on YouTube, in a day, which generate a large amount of data.

1.4.1 MARKETING

One of the primary links between social media and big data is marketing. Through obtaining data from social media such as Twitter, companies are able to "retarget," whereby the company tags online users when they surf on a certain website and more advertisement to the people who have shown some interests in the brand. Being able to retarget the potential customers is able to open up the market for the businesses as it is able to provide opportunities to aim for 98% of visits to retailers' websites which end up with purchases. Whilst it is common practices, it is understood that using big data on social media data on consumers is more sophisticated.

1.4.2 TRACKING ONLINE BEHAVIOR

Social media famous websites such as Facebook, Connect, and Google+ Sign-in, offer third-party declaration services which allows the user to sign up for an account with some correlated websites by using their social media account. These services allow specific users to browse behaviors that can be traced and tracked using software tracking devices called "cookies" and "pixels," which are available on computers when the user visits a particular website. The social media platforms are able to link the users' online activities outside of the social media platform to the data from their personal profiles. This allows other websites more access to the users' social media data. The excessive usage of social media has encouraged the development of new methodological and technical perspective to capture, process, and analyze big amount and complicated data. Analyzing the social media data can help one to improve the understanding behind how people act and think. As such, organizations are able to use these information obtained from the social media to think of ideas that could be applied to their activities, improve decision-making processes, provide services more effectively, able to have a target population for a certain innovated products, and also have the ability to influence consumers attitude in the next coming years.

In 2012, the United Kingdom Government had invested an amount of £189 million for doing the research in big data, moreover, an additional of £73 million was invested again which was released in February 2014. The economic and social research added another investment of the amount of £64 million in Big Data Network; this includes the funding to ease the access to social media data and further research leading to the total value of investment to £326 million by researching of big data alone. This proves that the access to such an amount of readily-available data on millions of people's' activities and behaviors are a highly useful resource for the researchers and also the organizations. To aid in data collection, a program has been developed that can automatically extract data from social media sites known as Application Programming Interfaces (APIs).

APIs are a set of guidelines and tools for sorting out interactions between different software, and their can also be used to extract social media data automatically. Some of the social media platforms are providing the APIs for free; however, there are restrictions on the amount or type of data that can be found by the users. Several companies, such as DataSift and Gnip, have dealt with social media companies to have more admission to large amount of data through APIs, such as the full public Twitter feed. After these processes, it is to sell to the third parties, with prices from around £2000 per month. Another benefit of having APIs is that it can also be used to enter some private data.

An example of tracking behavior and obtaining big data through social media is by using big data analysis. Big data analysis plays an important part in planning out the election strategy of a particular country. This campaign has hired a huge data analytics team; they make use of the data obtained from the social media. Through analyzing and browsing for correlations in past voter features and behavior, they were able to identify the profiles of the kinds of people who might vote for them, and to place the resources more efficiently. For example, television advertisements were broadcasted when it is known that they had most effects with the voters. Analytics was also used to identify which to target door-to-door.

1.5 MANAGING CHANGE TROUGH SOCIAL MEDIA AND BIG DATA

With the high traffic of social media, a huge cause for concern that plagues society, especially the generation Y, is that it can lead people spending far too much time on the social network. The social media applications such as Facebook serve nothing more than just being a form of entertainment value. They believe that it is a form of distraction and another thorn in academic achievements.

There are two basic issues that need to be pointed out in managing with big data, which are storage and management issues. First, how the data is being stored. The amount of data has drastically increased each time as new storage medium is invented. However, due to the huge usage of social media, there is no recent storage medium. Second, management is the most difficult issue to be solved with big data. This is because the way in order to resolve issues of access, metadata, utilization, updating, governance, and reference are proved to be critical barriers. Compare to the manual methods to collect the data, where the right protocols are often done step-by-step in order to ensure the validity, accuracy, and the digital data collection is much simpler. Data qualification is more focused on missing data or outliers than trying to validate every item. The way how data is collected, verified temporally and spatially when those data are readied for analysis and inspections (Kaisler et al, 2013).

1.5.1 INACCURACY OF INFORMATION

A study search habits of focusing largely on first year university students showed that, students relied on websites such as Google and Wikipedia about 80% of the data to obtain information. However, further research has shown that when asked to rank the reliability of information from different websites, it was found that Google and Wikipedia is the least reliable in providing accurate data. The reason for this could be that, due to the public having access and being able to edit information within these applications, they have the ability to change the information into random information that could be irrelevant to the topic and unless moderated properly, it could remain untouched and people may just believe in what they read. The educators have further critique social media for providing limited resources and useful information beyond just being an entertainment platform.

The researchers argued that lack of supervision and improper use of social media brought risks of privacy invasion and online harassment. The freedom and openness provided by such social media tools also lead to excessive sharing of information, yet, it is hard to prove if all the information shared is correct.

1.5.2 SOCIETY REDESIGN

A Study was conducted to answer a concerned mother opinion on social media on the Huffington Post, it was focusing on how social media is affecting teens in general and she outlined eight key problems with teens and social media. First, she believes that teens have lost the ability to pick up on social cues and that they are unable to read other people's' behavior which can lead to conflicts. Second, they are unable to learn the ability to cooperate with others which is essential in society to get along with friends and coworkers as they are mostly just spending time alone on the Internet. She believes that teens need to learn to be inclusive, learn how to deal with free times better, teens are becoming more sleep-deprived, create too much excessive drama on the social media, and are not appreciative of the present moments. The key highlight of the article is the final issue she brings up whereby she says that teens are becoming more aggressive and sexual when they feel anonymous and are communicating electronically. This means that they are able to hide their identity behind the technology such as creating fake accounts on social media which can be prone to cyber-bullying. However, she does believe that social media is fine but that teens need to be able to find a good balance between spending time on social media and spending time on real-life activities.

1.5.3 EDUCATION

While making use of social media in education, it can be seen as effective and efficient, it is worth noting that there are some key issues that may undermine social media in terms of education. With the advancement of technology, an individual may also need to keep up with the advancement in order to make full use of the technology. In order to make use of such technology, administrations and faculties staffs of higher education and even lower education institution are required to understand the phenomenon of social media technology. This can be a challenge as these staffs have to take out time from their busy schedule in order to learn about technologies and governments may have to invest more capital to make use of such technology and to provide courses for staffs to learn about the complex technology (Liu et al., 2018; Susanto, 2017a, Susanto 2017b, Leu et al., 2017).

The issues of the usage of social media in schools have different sides. While beneficial to teaching and learning, whereby the students are able to immediately and constantly connect and network understanding of their digital world. Guidelines and policies can be found on the Internet from various school boards and districts to make a distinction between purpose or intent and use of social media. These guidelines, however, are similar to those of that in an actual workplace rather than an actual policy document on social media tools. Issues such as security as well as facing the challenges of how and when it is an appropriate time for teachers and the students to engage socially online, as some may consider engaging socially online outside of school hours, but it can be considered as a form of harassment which can further lead to more issues such as sexual harassment as there have been cases reported worldwide of teachers and students engaging with each other outside of school hours on social media (Susanto and Almunawar, 2018; 2016; 2015).

However, not only the institutions should be aware of usage of social media. Students also have the same responsibility and should be aware that the comments and material which they share may have serious implications and even legal consequences for them and the institution. Students could misuse social media and post negative, damaging, and defaming material that can affect other students, staff, third parties, their courses, or the institution itself in general such as tweeting a complain about their homework or lecturer. Students may also use the social media to challenge the social norms and status quo and progress contentious, or provide controversial views and opinions. Social media works in a way that once postings have been made, it can spread very quickly to others and to remove the material or information can be very difficult and to handle the situation can be extremely difficult even if the authorities are involved. Additionally, misusing of social media may give rise to unfavorable publicity, which may result in damage to fame and reputation.

In such cases, institutions may also risk being exposed to complaints and will be accused of not having taken the appropriate precautions to prevent the inappropriate usage of social media by the students. For example, students posting inappropriate material can lead to complaints by other students, staff and third parties of bullying and harassment and discrimination. The victims of social media abuse can also accuse institution for having failed to properly take care of them and its obligations under health and safety rulings (Susanto, 2017c; Susanto and Chen 2017). However, it is well worth noting that having restrictions in place and taking action against the students, whilst monitoring their usage of social media having arisen debates regarding human rights and how it is taken away from them under the European Convention on Human Rights and is expected to further develop in the coming years.

1.5.4 CLAIMS OF ACCURACY CAN BE MISLEADING

Big data is understood to be able to offer a new approach to claim the status of quantitative science and objective method. In reality, however, working with big data can still be subjective and that what it is able to quantify does not necessarily have a solid claim of objective truth, particularly when considering the messages from social media sites. Bollier (2010) states that big data is not self-explanatory and that with such a huge amount of raw data and with it being so open to interpretation, he questions whether the data can actually represent an objective truth or that it can be free from bias and filtered properly.

1.5.5 ORGANIZATION IMPACT

Another issue of big data in a business aspect is that the consequence it has on the organization. With the advancement of technology and the usage of big data to analysis in most organization, it made the organization structure to deflect which means that the positions slot and roles within organizations had to be reelected and new positions are given. The major worrying issue here is that it might let the people to have lesser opportunity for veteran labors as well as technology replacing human labors. However, the data that were collected will require human to analyze, it requires a very different skills and approaches to deal with the data (Leu et al., 2015; Susanto et al., 2016b).

It has been argued that the size of the datasets has decreased, or even be removed, the need for created statistical methods such as random sampling, because all the data can be analyzed. However, in the case of social media data, it only contains data about people that use social media. There are concerns that social media data could not represent the groups in society, such as the elderly or those from lower income backgrounds. This means that there are large gaps in the data, and there are no acceptable methods for controlling for biases. There is also an argument as to whether social media data should be viewed and analyzed as quantitative or qualitative data. A number of researchers argued that more clarification in the methods used and consistency are needed for analyzing large quantities of unstructured and complex data (Susanto, 2016).

1.5.6 ETHICAL ISSUES

In the case of ethical issues surrounding big data came about in 2006, when a research group based in Harvard started gathering profiles of 1700 college-based Facebook users to study how their interests and friendships changed overtime. These data that were thought to be anonymous were then released to the world which allowed other researchers to explore and analyze the data freely. These researchers were then able to find out that they were able to de-anonymize parts of the data which would compromise the privacy of the students that were being researched with none of them being aware that their information were being collected. This was a major invasion of privacy and the case headlined and raised even more issues for researchers regarding big data.

1.5.7 ACCESS TO BIG DATA CREATES NEW DIGITAL DIVIDE

While it may seem that access to big data is straightforward, it is not as clear cut as it seems. The only social media companies truly have access to big social data and that only those researchers that are employed by companies such as Google and Facebook can gain access to such rich data when compared to just a normal scholarly researcher. Due to the data being such a privilege, some companies decide to keep the data for themselves and restrict access for others completely; some decide to sell the information for money while others decide to only offer a small amount of data for university researchers. This creates an unfairness in the system, whereby those that are employed by major companies or those that have a large amount of money are able to produce an entirely different data than the typical researchers that are not employed by the company.

The difficulty and the resources needed to obtain the big data produce a restricted culture of research findings. Big companies that hold a large amount of data are not responsible to make their data available to the public, and they hold the authority over who gets to see them. Big data researchers with access to proprietary data sets are less likely to choose questions that are contentious to a social media company if they think it may result in their access being cut. The chilling effects on the kinds of research questions that can be asked both in public or private are something we all need to consider when assessing the future of big data.

The digital divide then can be separated into two, the big data poor and the big data rich, with some researchers suggesting that with such a divide, they would rather not study cases relating to social media altogether as it creates a huge inequality and that the capitalists is very capable of manipulating the data provided to their advantage.

1.6 SOCIAL MEDIA AND BIG DATA LINKED TOGETHER

The information that social media applications contain is mostly valuable information that could be personal about individual. The data is valuable in the sense that it is detailed, personal and accurate as the information is provided by the individual themselves. The nature state of the data makes it vulnerable within a user's network. This can increase the possibility of accuracy when compared to the data from other sources. The data social media application stores are owned and controlled by private companies. Applications such as Facebook, LinkedIn, and the Google suite of products which includes but are not limited to Google Search, YouTube to name a few are driven by information sharing but are monetized through internal analysis of the data gathered, a form of computational social science. These data that have been gathered are usually used by business clients, government, other users within the social media platform, and also the platform provider itself.

Social media applications are capable of providing different variety of data. For example, when a user access the application, the time they are accessing can be seen and can then be used in different locations to identify insomniacs to advertise sleeping pills, or they could even analyze data on users' age and sexual preference to target advertisements of adult products. Such analysis, while capable of drawing in a bigger market for entrepreneurs, has the potential to cause unease or unauthorized disclosure of sensitive personal information about an individual. This poses both technical and ethical problems. It is possible to identify the individuals with a certain condition but whether they would like the information to be shared depends on the individuals themselves which brings about the major concern that it is unethical because social media takes advantage of our information for its own benefit (Oboler et al., 2012).

One of the major issues regarding big data in social media is the threat that it carries towards users' privacy from other users' media. It is a cause for concern and a huge threat due to the fact that the person harmed can theoretically not be involved in the uploading process of the data and thus is unable to take any precautions along with the amount of data, that is being uploaded to the database is so huge and varying that it cannot be manually sighted. Currently, to add more threat to the issue is that there is no countermeasures in place to protect against such a threat except post-priory legal ones where it is to prevent others from uploading potentially damaging content about someone. For this threat to take effect, there are two requirements that needs to be met. The first being, in order to cause harm to a person, a piece of media needs to be able to be associated or linked to the person is some form of way. The link can either be non-technical, such as being recognizable in a photo or technical such as a profile being hyperlinked to a photo. There is also the grey area of textual references to a person near to the photo or embedded in the metadata of said photo. This metadata does not directly create a technical link to a profile, however it is worth noting that it opens up the possibility for search engines to index said information and make it searchable, thus creating a technical link. An example of this is being able to reverse search an image on Google with a link provided and it is able to trace it back to its original source. The second requirement is a piece of media must contain harmful content for the person linked to it. This can again be non-technical such as being depicted in a compromising way. However, more interestingly it can also be technical. In these cases, metadata or associated data causes harm (Szonggot et al., 2012).

1.6.1 SOCIAL MEDIA

One of the ways to be able to cope with the usage of social media is by managing the usage of social media. With such a huge user base on social media, it should not be a surprise that almost everyone we know or are close with uses social media. It is important for institutions themselves to create effective policies and procedures and have them in place to address the issues that arise from the usage of social media, including the need to define the parameters of acceptable use.

For example, a number of institutions have been able to develop stand-alone social media policies which include appropriate references to other relevant policies and procedures. This can be seen in student codes of conduct and discipline guidelines such as school preventing students from bringing electronic gadgets to school or limiting their usage in class. IT and data protection policies where if students connect to the schools' Wi-Fi, they have limited access and all their data are secured and protected. Other institutions, however, have looked into updating their existing policies and procedures to address the issues of social media misuse. When creating or updating such guidelines, it is important for institutions to ensure that the acceptable behavior are clearly stated and briefed to students and that a well drafted procedure is in place to offer and effective framework to regulate the use of social media and through which to take action where issues arise.

Policies and procedures that are exceptional are also effective in assisting institutions in discharging the duty of care owed to staff and students and may support institutions in discharging the duty of under-equality and discrimination legislation by demonstrating their commitment to eliminating unlawful discrimination, harassment and victimization, advancing equality of opportunity and fostering good relations among groups. This basically means that, it works in a way that everyone feels that they are not being discriminated and that everyone has equal share of usage of social media.

Examples of comprehensively drafted social media policy should include several points, which are (1) define the true meaning of social media, (2) outline the acceptable usage of social media and define the behavior and conduct which is considered unacceptable, (3) state clearly whether the policy applies for educational and personal purposes or that it also affects usage outside the institution and student personal equipment, (4) outline the procedures of dealing with a breach in conduct and include references to other relevant policies and procedures.

1.6.2 BIG DATA

It is to analyze the amount of unstructured data produced by social media is difficult especially using traditional methods that rely on human analysts. That being said, social media analytics is a new field of study that is developing automated or semiautomated methods for analyzing data and, one such technique that has been developed is called sentiment analysis. Sentiment analysis makes use of normal language to process the techniques to read and understand the meaning to text, such as whether the author felt good, bad, or neutral. It is also able to provide wide insights on public reaction to a particular event in ways that was not previously possible. This method can be divided into two main methods which are the list or corpus approach and the machine-learning approach.

The list or corpus approach is the fastest and least labor-intensive approach, which means it requires the least manpower to operate. This approach is when a software is used to look for particular words, which can be considered to have positive and negative meanings.

The machine-learning approach builds on the previous approach by making use of machine-learning algorithms which involves a human analyst manually indicating how the program should interpret the use of specific terms or phrases in different contexts using examples of text. This requires more human input as compared to list approach but it does produce more accurate results as it is able to detect humor or irony, for example.

Sentiment analysis can be a useful tool to interpret public reaction to a particular event, such as a riot or TV show. Sentiment analysis is also used by traders to gauge sentiment between buyers and sellers to determine entry and exit points of a trade. However, at present, specific context of insights that can be concluded or drawn from such a technique is still limited and is still undergoing further research and development to improve upon the technology to apply it to wider settings (Susanto and Chen, 2018a; Susanto and Chen, 2018b).

An example of sentiment analysis is a project that is funded by the European Commission, known as "WeGov Project" which focuses on building the tools to analyze responses to government policies on social media. The development focuses on new technical aspects and methodological aspects to capture, process, and analyze large and complex data.

1.7 CONCLUSION

In conclusion, social media have derived benefits for users in variety of function, however, there are also negative impacts on social media. It is easy to access and useful for all users which make it convenient for them. It has been one of the challengers in the business market. Social media that is not found on Internet but available in other applications had become the most globally population social media platforms that provide various services. Furthermore, it is also convenient for student-learning purposes as well because information can be found through Internet and make it easier for them. In addition, social media allow business firms to interconnect with client without costing travel expenses as well as reduce time consuming. Brand loyalty is prominent for consumer due to help them in increasing revenues for firm, market share profitability and help in growth of firm which beneficial for firms. Big data is known as a huge and complicated data which are usually posted by big firms. There are three parts of big data such as volume, velocity, and variety. These instances are the way how data is collected, analyzed and also sorted in the company. Some data are openly posted to allow people to see while some data are confidential for their privacy purposes. These data are beneficial for some companies as they are able to track the usage of the consumer to gain more profits. How social media and big data are related is that they are linked in the market as company could advertise their products through the social media. Furthermore, it allows the company to track how consumer behaves by tracing cookies and also doing the big data analysis.

However, there are also some problems with social media is that data retrieved may not be fully relevant due to some human errors or it could be modify by nonexperts, for example, Wikipedia. Wikipedia does not provide relevant sources and can be changed in many ways as well as their meaning. In addition, social media in the organization has become very competitive because it required more users to sign up for it in order to be more profitable. In the society, the advancement of technology created an anti-socializing society among the teenagers with each other. Social media have been widely influenced by the mind of the society. There are also problems in big data as data may not be accurate or restrictions in obtaining the data due to access restrictions, for for example, requirement of registration fees or payment for downloading in order to obtain the data. On the other hand, it has been a concern to human that there will be an organizational changes as human labors could be replaced in the future. For example, the existence of robots in manufacturing companies will require less human labors. Big data and social media could have less security in some aspects that may cause ethical issues and also unsuitable to be on the Web. Social media may have a lack of security because some websites do not have age restrictions as well as leak of privacy information, therefore this can allow people with any range of ages to access to the data available in the social media which causes negative impacts for users. Moreover, cyberbullying could happen if the data uploaded lack the privacy. It would be easy for experts to track the system of the uploader in their own ways.

KEYWORDS

- **social media**
- **big data**
- **big data security**
- **business process reengineering**
- **core redesign**
- **information security**

REFERENCES

Acker, O.; Blockus, A.; Pötscher, F. Benefiting From Big Data: A New Approach for the Telecom Industry, 2013. http://www.strategyand.pwc.com/reports/benefiting-big-data.

Almunawar, M. N.; Anshari, M.; Susanto, H. Crafting Strategies for Sustainability: How Travel Agents Should React in Facing a Disintermediation. *Operational Research* **2013a,** *13*(3), 317–342.

Almunawar, M. N.; Susanto, H.; Anshari, M. A Cultural Transferability on IT Business Application: iReservation System. *J. Hospitality Tourism Technol.* **2013b,** *4*(2), 155–176.

Almunawar, M. N.; Anshari, M.; Susanto, H.; Chen, C. K. How People Choose and Use Their Smartphones. In *Management Strategies and Technology Fluidity in the Asian Business Sector;* Ordóñez de Pablos, P, Ed.; IGI Global, 2018a, pp 235–252.

Almunawar, M. N.; Anshari, M.; Susanto, H. Adopting Open Source Software in Smartphone Manufacturers' Open Innovation Strategy. In *Encyclopedia of Information Science and Technology, 4th ed.;* Khosrow-Pour, M., Ed.; IGI Global, 2018b, pp 7369–7381.

Almunawar, M. N.; Susanto, H.; Anshari, M. The Impact of Open Source Software on Smartphones Industry. In *Encyclopedia of Information Science and Technology,* 3rd ed.; Khosrow-Pour, M., Ed.; IGI Global, 2015, pp 5767–5776.

Bollier, D.; Firestone, C. M. *The Promise and Peril of Big Data* Aspen Institute, Communications and Society, Program: Washington, DC, 2010, pp 1–66.

Erdoğmuş, I. E.; Cicek, M. The Impact of Social Media Marketing on Brand Loyalty. *Proc.-Soc. Behav. Sci.* **2012,** *58,* 1353–1360.

Kaisler, S.; Armour, F.; Espinosa, J. A.; Money, W. Big Data: Issues and Challenges Moving Forward. In System Sciences (HICSS), 2013 46th Hawaii International Conference. *IEEE* **2013,** 995–1004.

Kaplan, A. M.; Haenlein, M. Users of the World, Unite! The Challenges and Opportunities of Social Media. *Bus. Horiz.* **2010,** *53*(1), 59–68.

Kirschner, P. A.; Karpinski, A. C.. Facebook® and Academic Performance. *Computers Human Behavior* **2010,** *26*(6), 1237–1245.

Leu, F. Y.; Ko, C. Y.; Lin, Y. C.; Susanto, H.; Yu, H. C. Fall Detection and Motion Classification by Using Decision Tree on Mobile Phone. In *Smart Sensors Networks;* Xhafa, F., Leu, F.-Y., Hung, L.-L., Eds., Academic Press, 2017, pp 205–237.

Leu, F. Y.; Liu, C. Y.; Liu, J. C.; Jiang, F. C.; Susanto, H. S-PMIPv6: An Intra-LMA Model for IPv6 Mobility. *J. Network Computer Applications* **2015,** *58,* 180–191.

Liu, J. C.; Leu, F. Y.; Lin, G. L.; Susanto, H. An MFCC-Based Text-Independent Speaker Identification System for Access Control. *Concurrency Computation: Practice Experience* **2018,** *30*(2), e4255.

Oboler, A.; Welsh, K.; Cruz, L. The Danger of Big Data: Social Media as Computational Social Science. *First Monday* **2012,** *17*(7).

Susanto, H. Cheminformatics—The Promising Future: Managing Change Of Approach Through ICT Emerging Technology. In *Applied Chemistry and Chemical Engineering, Volume 2: Principles, Methodology, and Evaluation Methods*; Haghi, A. K., Pogliani, L., Balkose, D., Mukbaniani, O. V., Mercader, A. G., Eds.; Apple Academic Press, 2017a, pp 313–332.

Susanto, H. Biochemistry Apps as Enabler of Compound and DNA Computational: Next-Generation Computing Technology. In *Applied Chemistry and Chemical Engineering, Volume 4: Experimental Techniques and Methodical Developments;* Haghi, A. K., Pogliani, L., Castro, E. A., Balköse, D., Mukbaniani, O. V., Chia C. H., Eds.; 2017b, Apple Academic Press, pp 181–202.

Susanto, H. Electronic Health System: Sensors Emerging and Intelligent Technology Approach. In *Smart Sensors Networks;* Xhafa, F., Leu, F.-Y., Hung, L.-L., Eds., Academic Press, 2017c, pp. 189–203.

Susanto, H. Managing the Role of IT and IS for Suppoting Business Process Reengineering. *J. Systems Information Technol.* **2016.**

Susanto, H. Smart Mobile Device Emerging Technologies: An Enabler to Health Monitoring System. *Kalman Filtering Techniques for Radar Tracking* **2018,** 241.

Susanto, H.; Almunawar, M. N. *Information Security Management Systems: A Novel Framework and Software as a Tool for Compliance with Information Security Standard.* CRC Press, 2018.

Susanto, H.; Almunawar, M. N. Managing Compliance with an Information Security Management Standard. In *Encyclopedia of Information Science and Technology,* 3rd ed.; Khosrow-Pour, M., Ed.; IGI Global, 2015, pp 1452–1463.

Susanto, H.; Almunawar, M. N. Security and Privacy Issues in Cloud-Based E-Government. In *Cloud Computing Technologies for Connected Government;* Mahmood, Z., Ed.; IGI Global, 2016, pp 292–321.

Susanto, H.; Chen, C. K. Information and Communication Emerging Technology: Making Sense of Healthcare Innovation. In *Internet of Things and Big Data Technologies for Next Generation Healthcare;* Bhatt, C., Dey, N., Ashour, A. S., Eds.; Cham: Springer, 2017, pp 229–250.

Susanto, H.; Chen, C. K. Macromolecules Visualization through Bioinformatics: an Emerging Tool of Informatics. *Appl. Phys. Chem. Multidiscip. Approaches* **2018a,** 383.

Susanto, H.; Chen, C. K. Informatics Approach and Its Impact for Bioscience: Making Sense of Innovation. *Appl. Phys. Chem. Multidiscip. Approaches* **2018b,** 407.

Susanto, H.; Chen, C. K.; Almunawar, M. N. Revealing Big Data Emerging Technology as Enabler of LMS Technologies Transferability. In *Internet of Things and Big Data Analytics Toward Next-Generation Intelligence;* Dey, N., Hassanien, A. E., Bhatt, C., Ashour, A. S., Satapathy, S. C., Eds.; Cham: Springer, 2018, pp 123–145.

Susanto, H.; Almunawar, M. N.; Leu, F. Y.; Chen, C. K. Android vs iOS or Others? SMD-OS Security Issues: Generation Y Perception. *Int. J. Technol. Diffusion* **2016a,** *7*(2), 1–18.

Susanto, H.; Kang, C.; Leu, F. Revealing the Role of ICT for Business Core Redesign, 2016b.

CHAPTER 2

IS IT POSSIBLE TO CONDUCT EDUCATION PROCESSES USING BIG DATA? CHALLENGES AND OPPORTUNITY

HERU SUSANTO

The Indonesian Institute of Sciences, Jakarta, Indonesia, E-mail: heru.susanto@lipi.go.id

Tunghai University, Taichung, Taiwan

ABSTRACT

Technology advances have eased the way technologies are being used, leading to more and more data being generated everyday through the usage of mobile phones, document uploads, e-mail and online chat that leaves data traces. This data that is so large and available everywhere, impractical to be managed with traditional analysis tools is known as big data. In recent years, with global changes, institutions of higher education have been changing the ways learning is being delivered compared to the past, as they are becoming more reliant on technologies. Big data has become more and more important not only for businesses but also for universities and higher educational institutes. In some universities, they have started working on data mining, extraction, and analytics to analyze the data that will benefit the field of education to adapt better to situations happening within and outside the learning environment and to optimize the student learning experiences. The main purpose of this study is to identify the ways in which big data can be used to support universities and higher education. There are also a few challenges of big data identified that need to be tackled before implementing the big data which could bring greater benefits to the universities and higher education, but at the same time could have other negative effects.

2.1 INTRODUCTION

With the increase in the number of institutes that provide higher education, competition increases as well as to which institute students get enrolled, which universities are popular and performing well in examinations. Decision-makers in those institutes need to choose their students wisely in order to maintain their high reputation and to produce graduates who are recognized internationally. Parents become more demanding when considering the universities to which they send their children, in terms of the facilities provided and valuable faculty members. Therefore, it has become a challenge for institutions of higher education to meet the demands of the parents and potential students due to global changes. During this time, big data becomes very helpful because it will aid decision-makers tremendously.

There was a lot of promotion of big data in universities in the year 2015. The growth of big data is not only useful in the organization but also in educational institutes. Big data is the origin from numbers extracted through conventionally-induced databases used in decision-making together with the new unstructured data. Some of the benefits of big data is that work is done faster and efficiently. Big data is useful in higher institutes not just to improve students academically but also to help improve the lecturers' performance and their efficiency by cutting down on the administrative workload. For example, a university can apply analytics to a certain large data by extracting useful data in distinction to large data. According to Philip Russom 2011)'s research report on *Big Data Analytics*, businesses are putting more effort into looking through big data uncover information that they did not notice in the past. He also mentioned that uncovering this information through big data is crucial now because of the economic inflation which has forced some organizations to go through deep adjustment, particularly those organizations that rely heavily on mass consumers. By using analytics, organizations are able to understand the conditions of the business now and at the same time still be able to trail the still developing aspects of their businesses such as, their customers behavior. Philip Russom defined Big Data Analytics as the progress analysis approach which functions on big data sets. There are three vectors of big data, that is, the volume, velocity and variety, and they have their own division for analytics. Holland (2015) defined the 3vs of big data and it is also explained below in the literature review as well as the discussion part of this report. Big data not only helps in the education sector but it can also help the organization to identify new business opportunities.

2.2 LITERATURE REVIEW

Today, in this modernized world, data is generated from everywhere, for example, through social network, sending of e-mail and receiving it, chatting with friends, uploading of documents and photos, making online purchases, online banking as well as online education. Most of this data is stored in the data warehouse which is also referred to as a centralized database. This huge and complex data is beyond the capabilities of the traditional methods to be analyzed and understood.(Manyika, 2011) Thankfully, with the recent developments in data mining and data analytics techniques, it is possible to collect and manage big data which can be used to turn this huge and complex data into meaningful information and patterns. In big data, there are the three Vs which are the key elements. First 'V' is variety which is the different forms of data available or the different formats of digital data. Second 'V' is volume—the scale of the data or how the units of data are being measured and lastly, velocity, which is the speed it takes to process the data or how fast the data is being produced. Big data, which may seem unimportant to all of us, is very important as it helps in decision-making. Big data, when analyzed, becomes useful information that is very valuable to decision-makers in all industries.

There are three stages to unlocking big data in any organization. Collection is the first step that should be taken to unlock the values of big data. This involves identifying data that can reveal valuable and useful information. The data should be checked for its sources to see whether it is relevant or not and stored in a form that can be used as it is not wise to invest in huge data and infrastructures if most of the data is seen as useless or irrelevant. Next is the analysis, once the data is proven to be useful, it has to be analyzed in order to get actionable information that can be used. However, it is a very complex process knowing that there is a growing diversity in the nature of data and analyzing these diverse data sets. Analysis has to have some form of linkage, connecting with different data sets to deliver the data that is intended so that people can have a grasp of that information. Thus, this is what the term complexity of big data refers to. Visualization and application is the last stage of data analysis. This is where the data is made available to users in a form that is understandable and interpretable so that it can be integrated into the existing processes and this can be used as the ultimate decision-making guideline.

According to Boyd and Crawford (2012), "the rise of big data that is a knowledge system is already changing the objects of knowledge and social theory in many different fields, while also having potential to transform the management decision-making theory." Various attempts have been made to utilize big data in education around the world. For instance, in places like the U. S, there are many schools that are making use of data-based decisions by utilizing the student's individual educational data to help them improve in their academic achievements. Big data can influence higher education practice by enhancing the students experience in order to improve the academic programming to a more effective-based decision-making and to a strategic response to changing global trends. For example, in the U. S, the results of the academic achievement of the student's data show that those who did well in Mathematics, which is one of the hardest subjects to most students influencing whether one is able to graduate or not, also perform well in their other subjects. Thanks to the advancement of information and communication technology, also commonly called ICT, there is a rapid growth in terms of the learning activities through the online system currently being utilized in places of higher education such as the U. S.

Big data is first stored in the form of large databases and then undergoes data processing. In big data processing there are actually four requirements in total. First requirement for big data processing is fast data loading and second is fast query processing. Third requirement is a highly efficient utilization of storage space and fourth is strong adaptivity to highly dynamic workload patterns. After processing, the next step will be big data analytics. Big data is important because it is able to provide information and valuable knowledge which enables one to make a better decision.

Big data needs to be analyzed using advanced analytical techniques so it is easier for people to have a better understanding. This is where learning analytics comes into play. "Big data incorporates the emergent research field of learning analytics which is already a growing area in education. (Long and Siemen, 2011)" However as stated, the research in learning analytics is only limited to the examining indicators of class performance and individual students. Changes of big data analytics depend on firstly, data storage and architecture, secondly, data analytics processing, and thirdly, data analyses. The field of learning analytics is currently expanding with the widespread big data availability. Learning analytics can present the area of interest, education curricular as well as the future career for a specific student, and may provide an individualized learningbecause it can actually be used to create individually customized learning methods and information, according to the needs of each student (Kim and Ahn, 2016). With the rising number of meaningful online data, there are statistics which show that their usage is minimal. This is due to the unawareness of the users on how useful the data can be, as it is just chunks of data without meaningful information, if no further analysis is done. Thus, many countries are starting to realize that and are investing a lot of time and effort in developing new infrastructures to help analyze big data by gathering information from various study methods and producing meaningful results by using the data available inside and outside of the schools.

The most important technology in utilizing big data is the data mining technique. This technique is used to extract meaningful information from the vast amount of data in the field of education (Linoff and Berry, 2011). This meaningful information can be very useful in providing individual learning experiences for the students, given the fact that it can create customized learning methods to cater for individuals. For example, analyzing student logs using the mining technique can help lecturers get valuable insights from the data analysis to help them understand the behavior of different students better and with that they can customize education services that are suitable for every single student. It is also important to constantly keep track of the progress of each student as this real time data can be used to be analyzed further for future improvements.

Long and Siemen (2011) defined learning analytics as "measuring, collecting, analyzing, and reporting the data for learners and their contexts with the purpose of understanding and optimization of learning and the environment in which it occurs." According to Simen's United Nations Educational, Scientific, and Cultural Organization (UNESCO) policy brief, there are major application plans for learning analytics which include learning analytics dashboard, prediction analysis, adaptive learning analysis, social network analysis, and discourse analysis. The function of the learning analytics dashboard is to visualize the log data in the form of tables and graphs to make it simpler for individual users or groups to understand them. For example, students can use their own log data in comparison with those of others to find out where they are standing, whether they are on par with them or behind them. Predictive analytics is used to predict students' academic performance and grades. For example, it can also be used to alert learners if they are moving towards the danger steps with the help of predictive analysis and could give them an alternative course of action to guide their path entering the average or excellent step.

Adaptive analytics helps to assess a student's level and attention to a particular subject. For example, it will provide students with directed learning to enhance the results of the study by providing the necessary skills and information to help in the understanding of a particular subject. Social network analytics help to reflect the human relationship. For example, this relationship can be formed in most schools when the student interacts with the learning management system through e-mail and discussions where information is exchanged. Discourse analytics analyze contents, comments or thinking, text of students which is particularly useful to measure the quality of online education. Text mining is a part of discourse analytics. For example, online text chat can be analyzed to find out the degree of interactions between the students, and measuring and assessing the lessons to see the progress of the students.

More and more data mining techniques are becoming available to extract complex data turning it into meaningful information that can be used in practice. This creates new possibilities of 'big data and analytics' which is currently happening in the field of education.(Harper and Parkerson, 2015) At this early stage, most of the work on analytics within higher education comes from interdisciplinary research, spanning the fields of educational technology. The core element in analytics in education is centered on data mining. Big data has become more and more important for educators and learners because of the potential to revolutionize the ways research and practice are carried out. Big data and analytics can be used to support education in several ways, for instance, to improve the decision-making where critical because the data taken from the students is real time. The more structured and unstructured data to be accessed and analyzed is, the more sophisticated decision-making becomes. This will in turn lead to better performance, reduced risk, and improved efficiency. Therefore, the analysis of big data is very important. Mayer-Schonberger and Cukier, (2013) sum up the potential of big data in one succinct sentence: "things that can be done on a large scale that cannot be done on a smaller scale to extract new insights or create new forms of value."

Goes (2014) presents "a taxonomy to describe big data analytics research projects." Decision time is very critical as student data is generated real time, for example, from the real time online learning activities to weekly, such as their class attendance and yearly, such as reports on their progression statistics. Big data analytics simply refer to the "usage of data, statistical analysis, and explanatory and predictive models that will help them to gain insights and to present that data through various visualization forms". The goal of visualization is just for users to have a better understanding so they can interpret large and complex data sets. Visualizations could be from a range of simple pie charts to more complex visualization forms such as an interactive tree map. The techniques refer to the use of statistics, advanced computational and machine learning method to help analyze or predict trends in data. This is basically how big data analytics works and is used in decision-making for higher education.

Big data connotes the interpretation of a wide range of data which when processed can help to predict the future performance of a learning institution and identify potential issues related to academic programming, research, teaching, and learning. Institutional effectiveness and adaptation to change relies on the analysis of appropriate data. Big data not only enhances a student's learning experience and improves academic programming but also leads to a more effective evidence-based decision-making and responds strategically to global changing trends. Hilbert (2013) mentioned big data as a cost-effective prospect to improve decision-making. Institutional analytics and information technology (IT) analytics play a major role in the processing of big data in institutions of higher education. Institutional analytics enables an institution to make timely data-driven decisions across all departments and divisions. On the other hand, IT analytics mainly deals with student information, learning management, and learning experiences. Besides these two, there are also academic analytics and learning analytics which are crucial to any higher educational institution.

Big data is starting to grow exponentially especially in the field of education. According to Romero and Ventura (2010), the governments are beginning to see the potential of big data for education as they have seen what it can do in the areas of education. It can be utilized to help evaluate the performance of both the schools and teachers, support the competitiveness, and also improve the efficiency. Hilbert (2013) pointed out that big data delivers a cost-effective prospect to be implemented and is efficient as the use of data can help to improve the education and decision-making bringing it to the next level. In higher education, there is also an increase in the usage of data mining and predictive analytics which can be used to predict the student behavior and classify them. In some cases, it can also be used in monitoring the student progress, recommend suitable course for individual needs and to even create a personalized network for students. Predictive analytics can be used to help predict the number of students likely to dropout from schools and how many students are unlikely to make it through with passing grades. In addition, by capturing and analyzing this data, it can be used in the research and practice that will have positive effects in the field of education to help reduce the number of students that are likely to dropout from school and support the teachers to help them better understand their students individually, and come up with a course of action to reduce the impact of the predicted results.

As big data evolved over time, new software approaches also known as analytics are now used to help in extracting valuable information from the big data, which could be used in higher education where lecturers will be able to access large quantities of complex data that is transformed into very meaningful easily understood information for example, accessing the grades of students in each course over a period of 1–2 years. This information can then be used by researchers to study the patterns of the student's performance over a period of time. With big data, every single student that registers in the college could be checked in terms of their course assessment, their web activities, where they gather their research documents and valuable information from and even the patterns generated from their exams on how they answer every question. In addition, analytics software can be used to assist in the analysis to come up with a course of action with the data collected.

According to Picciano (2012), analytics refers to use of data mining techniques to develop decision-making process that "converts data into actionable insights, uncover patterns, alert and respond to issues and concerns, and plan for the future." Big data analytics can also be useful in terms of monitoring student's performance, supporting decision-making and the student's admission process which will help shorten the lengthy process. It was also mentioned that colleges can benefit from the use of big data and learning analytics by evolving themselves from the traditional methods of how instructions are being delivered face-to-face to a more time-sensitive learning analytics application for example, by utilizing the learning management system (LMS), which is commonly used in universities to provide easy access to lecture slides, discussions in discussion board, doing quizzes online, and to monitor the student's activity.

Without data mining techniques to make sense of the bulk data or complex data also known as big data, this data will be more or less useless as it is just data that does not have meaning to it. With time, more techniques or algorithms have been developed to help extract this data, building some sort of pattern that could be easily understood and be used. Big data has been found to be useful in the field of education and is seen to be widely implemented in institutions of higher education. Data mining tools such as the Moodle which is known for its open-source learning platform used in universities, Meerkat-ED known for its web tool used to analyze interactions of participants in the forums, and Graphical interactive student monitoring tool (GISMO) which is an interactive tracking system built for Moodle are some types of learning analytics (Reyes, 2015). Learning analytics like these can be used to determine for example, the probability of a student being successful by looking at their past performance and perhaps accessing the demographic outlook to see whether students with marital status are most likely to achieve a better grade point average (GPA) compared to others or not.

As stated by Vincent (2016), in the field of higher education that has implemented big data, higher educational institutions are beginning to reap the tangible and intangible benefits that come from it. For example, in a latest UK-wide study, it shows that in the year of 2011–12, the universities generated £73 billion in output as compared to previous years with only £59 billion. This shows that there was an increase in output of up to 24%. The reason being, higher educational institutes are now making critical and strategic decisions with the support of big data as it gives valuable insights gained from data collecting and analysis. With data mining tools, universities have been putting a lot of effort in compiling and tracking the student's data more than before, for example, from students entering and leaving the university, student performance details, learning progress, and their probability of success. The use of big data in the field of education will lead to a better quality of education and more experienced students and teachers as compared to the past.

2.3 RESULT AND DISCUSSION

As mentioned above in the introduction and literature review in this report, big data consists of three vectors that is, the velocity, volume, and variety. They characterized in what way big data varies in distinction to the traditional way of managing data. The three vectors of big data were introduced and acknowledged on the 6th February, 2001 by Doug Laney (Borne, 2014). Below is the explanation of the three vectors of big data.

The first vector is volume, where it means when an institution or management gathers data through a series of origins which includes gathering data through social media or through interactions between business and customers. The collection of data was previously a problem, however, recent findings such as, Hadoop has made storing and collection of data much easier than before. The volume of big data refers to the volume of information in order to achieve relatively unimaginable areas. According to Soubra (2012), the volume of accessible information is increasing towards a rising percentage currently. It does not limit to just one person but also to organizations. The beginning of information will be combined at a constant interval. Formerly, the entire organization's information and data was developed by the staff whereas presently, information and data is developed by three stakeholders of the company namely the consumers, staff, and lastly all the associates. However, for a corporate group, the information is developed with the help of a machine. Soubra went on giving an example where he mentioned that there are more than hundreds of millions of mobile telephones which transmit a diversity of information to the hardware and software of a whole network. He said that this information was formerly not in existence (Almunawar et al., 2018a; 2018b; Susanto et al., 2016; Almunawar et al., 2015).

The second vector is velocity where information torrent at an extraordinary pace is handled conveniently. Gewirtz (2016) defined velocity as the frequency of how quickly the information is received. Soubra mentioned in the beginning, organizations evaluate information with the help of batch processes whereby an individual receives a lump of information and then surrenders the task to the program and holds up for transmission of outcomes. It would operate only if the approaching information percentage is behind schedule than the batch processes and also when the outcome is beneficial even though it is slow. However, amidst modern origins of information in the world today such as phone applications, these batch processes could collapse. The information torrents towards the program in actual time and is only significant when the setback is brief.

The third vector is variety where information is of both structured type such as binary data and unstructured type such as audio and e-mail. Gewirtz takes e-mail messaging as his example for variety. He mentioned that the process needs to be examined among the collected and selected messages and they contained the sender e-mail address, time and the end receiver of the e-mail, and each email has *"human-written test and attachments."* Rearranging the information significantly is quite difficult especially because it keeps changing.

An article written by Rijmenam (2013) suggested four other vectors which can help people comprehend the amazingly complicated description regarding big data. He mentioned that these four vectors are also crucial in establishing big data approach. The other four vectors are veracity, variability, visualization, and value.

The additional first vector mentioned by Rijmenam is veracity and he states that it is useless if the information coming is rapid in quantity but is not correct. This imprecise information could lead to loads of complications for companies as well as customers, hence the companies should make sure that information and the analysis of the information is precise. He also states that the information and analysis needs to be accurate specifically in computerized decision-making in which the individual is no longer needed. To conclude for veracity, it simply means verifying so that the information is always correct.

Next vector is variability. Rijmenam said that variety is usually mixed up with variability by most people. An example he gave comparing both variety and variability is that if an individual sold various buns then it is variety, but if a customer frequents the bakery every few days in a row and has gotten the same buns everyday but the same buns do not have a consistent smell and taste, then that is variability, he explained. Variability gives a meaning of constant modification. To conduct a suitable opinion analysis, it is important that the text is understood and interpretation of the accurate definition is translated which is close to being problematic. On the occasion that the definition is consistently uncertain, this could create a large shock on the data adjustments.

Third vector is visualization where it is the most difficult component of big data. Visualization included complicated diagrams which consist of diverse changeable information and at the same time are able to remain comprehensible and justifiable. This vector is not the ultimate greatest mechanical problematic section. However, to tell a complicated description from a diagram would be very problematic and excessively critical. Rijmenam states that fortunately to a greater extent, big data foundation has appeared and they target these sections and visualization could finally build a variation. In the subsequent years to come, this could be a path to follow in which visualization can aid the company to justify inquiry that the company would not realize they could ask. It might be better and more productive to envision a huge amount of complicated information with the help of graphs and diagrams to deliver interpretation rather than using reports and spreadsheets.

The last vector introduced by Rijmenam in the article written by him, in the year 2013, is value. The value in the analysis conducted on the data which in turn shows that information could evolve into knowledge. Value in a company means how the companies would use the information and change their company to an "*information centric*" one which depends on observations retrieved from the analysis of data during the managerial process.(Rijmenam, 2013) It is important to ensure that a company is gaining value from the information after acknowledging the other vectors of big data.

According to the article posted by Rossi in 2018, he identified four ways in which we could incorporate big data to education. He states that universities have information overflowing in distinction to every angle through the social media, surveys and online activities such as lecture exercises and quiz assessment. Any university or academy which is unsuccessful to grasp big data alternatives would not be able to catch up to those universities that use big data. Therefore, Rossi has suggested four ways as to how a university or higher academy can implement big data awareness.

First is by starting small, where it is necessary to uncover the scope of operations which is the maximum comprehensive data, for example, function that can pinpoint the complication which could be resolved easily with analysis. The strength from fast achievement would enable the management to solve more complications all over the universities.

Second is by analyzing the wanted results. Rossi starts this section by stating that in order to have a favorable analysis route, it is important to start off with the conclusion subconsciously by asking questions such as how can you improve enrollments, is the university targeting to boost connection and communications between students, and also how and through which ways can the universities get income. It is important and wise to also conclude the wanted results of the universities and after that advance the strategy and information construction which can aid in achieving the university's goals before devoting money and time to any particular big data analysis actions.

The third way suggested by Rossi, is to promote data culture. It has been a while since universities and higher academies have had any revenue to keep the information, let alone examine it. With the help of Hadoop, it is not a problem anymore. Rossi mentioned that it is costlier to dispose it rather than to preserve it. It is necessary and compulsory to develop a culture which regards information as important and at the same time creating data analysis a necessary part of any companies' agreement.

The last way suggested by Rossi is finding reliable partners. He suggested that finding a colleague that is devoted and hardworking can help the companies to succeed in all processes of the analytics adventure. They can now aid in grasping the investments and help in engaging steps which would not need difficult to discover sets of skills that have the prospect of turning obsolete and old-fashioned in a few months' time. Through the use of new and improved implementation and learning, higher educational institutes and academies could upgrade their development and appeal to the younger generation which can differentiate them from others. Rossi concludes by saying that we do not have to review big data as a dilemma but we could utilize it to our benefit by altering our access.

It is often made up by individuals that big data methods are not prejudiced because of the proportion of the data and the methods that are introduced through algorithmic systems. Nevertheless, it is not correct to conclude that big data are objectives just because they are data-driven (Muñoz et al. 2016). They also state that the challenges of developing honesty and reducing the prejudiced effects of data which can be categorized into:

I. Challenges relating to data used as inputs to an algorithm

The algorithmic systems of big data hire cultured techniques in which inputs are needed. A few of the technical themes which could lead to prejudiced outcomes are such as:

- Poorly selected data
- Incorrect, outdated or incomplete data

- Selection bias
- Unintentional perpetuation and promotion of historical biases

II. Challenges related to the inner workings of the algorithm itself

The technical means included in algorithmic systems are generally undiscovered to some potential students, consumers, or the public as they are mostly considered as confidential to the individual that benefited from them. Without transparency and accountability, there might be flaws in using big data such as:

- Poorly designed matching systems
- Personalization and recommendation services that narrow user options instead of expanding them
- Decision-making systems that assume correlation necessarily implies causation
- Data sets that lack information or disproportionately represent certain populations

One of the main reasons how big data can support a higher institution or university is through engagement between the students and the institutions. According to (Schmarzo, 2014), he mentioned that universities or higher educational institutes have introduced a large number of students' engagement points from big data which can help the students and it starts from initial profiling all the way till they graduate from the institutes. He defined student engagement life cycle as how the institutes influence and improve the student's engagement and also how it could increase their time in the university. In that very same article, Bill Schmarzo then listed nine big data-powered applications that could be beneficial for higher education. The nine big data powered applications are as follows:

i. Student acquisition

Under this section, he mentioned the use of information from older performances and also data of both current and past students, and then by using the information to come up with profiles that can most likely tell them what kinds of students are likely to be registered in those particular institutions. He also mentioned the need to employ graphic analysis to review their student's social network to pinpoint their friends or acquaintances who could be the institution's potential new scholar.

ii. Student course major selection

The universities or institution could use the student's previous data such as their high school results and performance, and also some other test that they did such as the aptitude test, surveys about their interest and correlate those results with the university graduates in which they can advise the students which major to take or what curriculum to take. It is also necessary to incorporate outsourced data in regards to the future workforce competence, ability requirements and income which can help the students to make the correct choice to decide which minor or major will be suitable for their future.

iii. Student performance effectiveness

They should also keep track of the students' test results and compare those results to their previous ones and group similar students with similar grades together. The lecturer's notes and social media data such as tweets, blogs, YouTube and hashtags should be incorporated in order to come up with a much more precise profile of the individual's weakness and point of views. Creating individual tutoring or small group tutoring for those students who have problems or even advising them to change their major could be recommended to help those students who have trouble coping up.

iv. Student work groups

To improve the student's individual performance, the universities can arrange for a leverage cohort analysis, which enables the students to work together both inside the class and also outside the lectures that can help them to improve individual achievements. This activity then enables the lecturers to identify the group's assignment, the reasons and factors for those assignments, and it also allows the lecturers to cancel. These analytics can reorganize the group task depending on the outlined design components and also other factors in which the lecturers need to write down their conclusion that are to be combined with the purpose of the group performance after reorganizing in order to amend the data set.

v. Student retention

Next application is to incorporate the past data and grades which include their effectiveness and also their group works where it then connects with the respective demographic, social, and also financial data to:

- Grade the attrition possibility, and
- Distribute suggestion which grants the universities to come up with a conclusion either to keep the student or not

By distributing and measuring the success of a particular suggestion, it allows the lecturers to come up with their own suggestions which could be checked for outcomes and can be practiced in the forthcoming detention interference propositions.

vi. Teacher effectiveness

This application section concerns in adjusting and measuring the lecturer's achievements. Their achievement can be measured by the number of students, the students' probabilities and their attitude categories, subject matter, and also a few other variables which can certify that the lecturers have the top experience for the student and lectures.

vii. Student lifetime value or booster effectiveness

It is necessary to plan in advance with appreciation offering to possible levels for both the undergraduates and postgraduates. Major influence in targeting, profiling and messaging to improve postgraduate giving are by understanding the possibilities to propose now or forthcoming potential of income and prosperity. Individuals who take advantage of these are the initial recognition of forthcoming supporters.

viii. Student advocacy.

Data from the individual's social network or their grades are used to come up with a student advocacy score which can influence the student's:

- Acquisition in which it targets a happy student's colleague
- Retention by flagging any adjustment which could be a forerunner to retention complications

- Performance effectiveness by flagging adjustments which could be forerunners for lecture room achievements complications,
- Lifetime value apps

ix. Bookstore effectiveness

The last big data-powered application is the bookstore effectiveness where it is the use of retail industry's finest proceeding to upgrade bookstore benefits using analytics-driven applications such as textbook stock optimization and also the merchandising effectiveness.

As mentioned in the literature review above, the significant technology used to extract useful information is data mining. There are two areas that use big data in the education sector which are educational data mining and learning analytics. In general, educational data mining searches in the most recent design between data and then cultivates fresh miniatures whereas learning analytics implements foreboding miniatures in teaching organizations. According to a report written by Bienkowski et al. (2012), it is mentioned that both educational learning analytics and data mining utilize constructing a miniature together with analysis which could impact the online learning systems. Educational learning analytics and data mining continue to cause visible data that has previously gone undiscovered hence no action has been taken. They also mentioned in their report that in order to help add the fields and improve value from their rational functions, they recommend lecturers and authority to:

- Cultivate experience concerning utilizing data for generating teaching choices
- Associate the IT division in outlining for data grouping usage
- Begin with a concentrated scope where data might assist, display success, and then widen to new scope
- Communicate between the scholar and parents about the location of where the data approach from and how it is utilized,
- Aid in coordinating state procedures alongside technical requisites for online education structure

Educational data mining analysts for example, Baker and Yacef (2009) and Baker (2011) viewed those mentioned below in act of the objectives as part of their analysis:

- Creating student's models that include their precise data for example, their knowledge and attitudes in order to predict their forthcoming learning
- Reveal and reconstructing domain models which define the content expected to grasp and the best education classification
- Learning the effects of several types of teachings possibly contribute through a learning software
- Building a computational model that can mix the models of the domain, students and software culture in which it can advance systematic philosophy between learners and learning.

To accomplish the goals mentioned above, Baker uses five categories. The five categories are described below:

1. Prediction

First category is prediction where it requires creating a model that could derive a single form of data from the merger of other forms of data. Examples of using prediction can include uncovering the student's attitudes when they could not answer a question correctly even though they have the skill to and when they are not taking the system seriously. Predictive models have been utilized by many universities to get to know what kind of attitudes the students put up with when they are in some online study surroundings, which includes answering practice tests and also when they are engaging in discussion forum.

2. Clustering

Second category is clustering where it means the discovery of data points that are generally categorized together and could help to divide an adequate collection of data into divisions. An example about clustering applications are categorizing students according to their study arrangements.

3. Relationship mining

Next category is relationship mining, where it includes uncovering connection among variables amid files as well as encrypting the variables in the act for consequential purpose. An example is that it is able to recognize the connection among the alliance amongst products bought from an online shopping.(Romero and Ventura, 2010) Marceron and Yacef (2010) mention that association rule mining could be utilized to find the student's error that occurs more than once by linking the information between the user class to create suggestions for content which is most probably useful or for making adjustments to teaching access.

4. Distillation for human judgment

Next category is distillation for human judgment, where this is a method that associates representing data in a way that allows an individual to recognize instantly or categorize characteristics of the data. This upgraded the machine-learning models because individuals can pinpoint the patterns and characteristics of the students' attitudes and their learning patterns.

According to the report written by Gupta et al. (2014), although data mining is one of the most important technologies as mentioned in the literature review above, there are a few challenges faced when enforcing the process for interpreting the trend the institution had. Some of the challenges faced were:

- Data fog situation such as the certainty of the data
- Finding the relevant and accurate dataset for analysis
- Grouping the dataset based on the requisite of the program
- Inadequacy of data governance
- Getting to know the algorithms given by the add-in

Bienkowski et al. (2012) mentioned that learning analytics point out data collection as well as the measurement for activities which universities require to engage in as well as to comprehend and at the same time balancing the reasoning and broadcasting of data. They also mentioned within their report that learning analytics allows human tailoring of responses for example, giving feedbacks and customized teaching content. Learning analysts could hire:

- Social network analysis for example, student to teacher relationship to come across disconnected students
- Data that consists of giving information regarding another data which can regulate what end users interact with

According to the report written by Strausser (2015), one of the reasons to utilize big data analytics throughout universities is mainly to boost retention percentage between students. He mentioned that most of the institutions and universities are being assessed with the use of profitability measures and business success because they are often faced with repeated financial complications. Strausser states that it is important that both the higher institutions and universities review and analyze steps to be done in order to raise retention. Susanto et al. (2018) defines analytics as a process of establishing awareness over problem explanation. The knowledge of analytics and big data used for universities has expanded over the years because of the decline in capitalization and poor weather. Universities and higher institutions are noted with compiling extensive data for example, student's analysis outcome as well as the graduation rates (Susanto, 2016; Susanto et al., 2016b). However, evaluating this data to make a judgment and modification was not very favorable. Strausser further mentioned that universities have joined big data analytics with its processes into three phases. The three phases are:

1. Learning management system, for example, blackboard is used either in the old ways in classrooms or used in conferences, sometimes
2. Web 2.0 technologies, for example, social networking
3. Analytics whereby they are being included currently

Learning analytics is based on interpreting and gathering data from a few different origins to contribute knowledge to what can help or cannot help both the education and teaching aspects. The work of Buckingham Shum (as cited in Strausser, 2016) mentioned that there are three parts to learning analytics. The three parts are macro level, micro level, and lastly meso Level. The first level, macro is that it allows the distribution of information through different universities and institutions in which they are most of the time described as best practices. Second part to learning analytics is the micro level and, for instance, dashboards are usually incorporated into the learning management system. Lastly is the meso level where it helps in the decision-making of the universities or institutions. Postgraduates boost the universities' or institution's reputation and they also help by boosting the number of admissions. Strausser states that universities have been surveying outlines through Amazon, Netflix, and also Pandora. Their aim is to utilize common characteristics of the analytics tools for example, the registration systems. He went on giving an example of big data analysis in the education sector, for instance, Santa Monica College's glass classroom initiative in which the student's achievements would be assessed through gathering a certain amount of data. Through the student's assessment, the program is able to give more needed examples and practices which intend to make sure that the goals are achieved. Another example he gave is that of the University of Wisconsin-Madison, where engineers are coming up with an initial cautionary structure for the lecturer in which they will be notified about the possible risks for the student. The goal of this system is to detect at the initial stage and to boost the student's progress which then helps in the universities and institutions retention and graduation percentage. The above mentioned tools are able to help lecturers at both the institutions and also in the learning management system (Susanto, 2018; Susanto 2017c; Susanto and Chen, 2017, Leu et al., 2015; 2017; Liu et al, 2018).

2.4 CONCLUSION AND RECOMMENDATION

In conclusion, big data is growing in importance as it can bring a lot of valuable insights to different fields of work including the field of education. Big data can also be used to address the challenges of finding information at the right time when the data is scattered across several unlinked different data systems in institutions. By identifying ways of aggregating data across systems, big data can help improve decision making. Analytics has the strong potential to support learners and lecturers to recognize when there are danger signs before it threatens the learning success. However, wide institutional acceptance should have a clear strategy and the usability of the analytics software packages. Big data can support the learning needs of individual student. For example, using learning analytics as a fundamental part of big data in universities and higher educational institutes can provide researchers with opportunities to carry out real time analysis of learning activities. Despite the uncertainties, the growth of learning analytics simply means that we are not only considering the opportunities offered for more effective decision-making for higher education, but we also have to look into the ethical challenges in institutionalizing learning analytics to drive and shape student support. By analyzing the student's data, predictive models can be created to identify the risks and provide appropriate interventions to counter it. This will enable instructors to adapt to their teaching, tailored assignments, and continuous assessment.

In the field of education, big data can be used to guide students to decide which major course to take that is most suitable to their needs and closer to their interest by using the data generated from their previous school years and results. Taking the right major is very important as some students might decide to drop out after spending a few years on that major after realizing that their interest is not in that area, which is really sad to hear and a waste of their time as well as money. With the support of big data, without doubt positive impact can be brought about to help boost the student's performance effectiveness. This can be done by analyzing their data in comparison with other students who are at the same academic level to find out the different alternatives on what they can do and provide the steps for them to follow so they can perform more effectively. Lastly, it can help to analyze learning patterns, providing guidance for the lecturer on how best to deliver the course to their students, in which they will be able to have a better learning experience (Susanto and Chen, 2018a; 2018b; Susanto, 2017a; 2017b).

There are a few reasons why higher education institutes should adopt analytics. Some of them are to increase financial efficiency, to expand their impact both locally and internationally as well as to respond to the demands for greater accountability to their stakeholders and clients. Advantages of analytics are that it serves as a foundation for making wiser choices, reducing redundancies, and saves a lot of time. In other words, data warehousing can be an effective method to unleash the usefulness and potential of big data in higher education. Besides this, data warehousing implements standardized data formats which results in more accurate data representation. This in turn prevents confusion when data is being shared among the different departments. Last but not least, a data warehouse has a huge storage capacity for historical data that can be readily used for experimentation. Historical data can then be analyzed according to its time line and trends which help to predict the most likely future trend (Almunawar et al., 2013a; 2013b).

However, the biggest challenge here is how to make sure that we have enough skills to handle the big data to make full use of this technology at our disposal and giving meaning to the data collected. There are also other challenges such as privacy and security issues, people are becoming insecure with all this sharing of data that can be widely accessed (Susanto and Almunawar, 2018). Even though it is for a good cause, it can also bring harm to them if the data is misused or hacked. If these challenges can be overcome, it will really have a positive impact not only on the society but also on the areas of education, where learning can be brought to another level, where it was once impossible in the past but highly possible now with the advancement of technologies.

In our opinion, there should be a collaboration between the learning institutions and big data integration to have successful implementation of big data as it is essential for them if they plan to pursue the goal of comparable and sharable data. As most of the data is scattered across departments and comes in different formats, it makes it difficult to be consolidated. They need to have a certain transparency towards one another so that they can effectively utilize this data to work better for the benefit of both parties. Collaboration may involve for instance, joint projects where both parties can involve their IT specialists and teachers of higher education. By combining data mining, programming, learning analytics expertise with higher education, the capacity for future alliances could be built between them.

There should also be a contingency plan to anticipate when there is a pitfall of data-generated fallacies of predictive analytics that will be very crucial to the progression of big data. The inherent concerned needs to do something about their data security such as boosting up their Internet firewall system to protect their data, routine checks on their data encryption and rules around data management and collection should be stated clearly. In the future, there should be identification and establishment of policies to specify who is accountable for which portion of the institutional data and information, accessibility, consistency, maintenance, and authorization if there is a need for data transferring or accessing the restricted data. All these policies should be stated clearly in order to avoid unnecessary mistakes (Susanto and Almunawar, 2016; 2013).

KEYWORDS

- **big data**
- **data mining**
- **learning analytics**
- **learning management system**
- **visualization**

REFERENCES

Almunawar, M. N.; Anshari, M.; Susanto, H. Crafting Strategies for Sustainability: How Travel Agents Should React in Facing a Disintermediation. *Operational Research* **2013a,** *13*(3), 317-342.

Almunawar, M. N.; Susanto, H.; Anshari, M. A Cultural Transferability on IT Business Application: iReservation System. *J. Hospitality Tourism Technol.* **2013b,** *4*(2), 155–176.

Almunawar, M. N.; Susanto, H.; Anshari, M. The Impact of Open Source Software on Smartphones Industry. In *Encyclopedia of Information Science and Technology,* 3rd ed.; Khosrow-Pour, M., Ed.; IGI Global, 2015, pp 5767–5776.

Almunawar, M. N.; Anshari, M.; Susanto, H.; Chen, C. K. How People Choose and Use Their Smartphones. In *Management Strategies and Technology Fluidity in the Asian Business Sector;* Ordonez de Pablos, P., Ed.; IGI Global, 2018a, pp 235–252.

Almunawar, M. N.; Anshari, M.; Susanto, H. Adopting Open Source Software in Smartphone Manufacturers' Open Innovation Strategy. In *Encyclopedia of Information Science and Technology 4th ed.;* Khosrow-Pour, M., Ed.; IGI Global, 2018b, pp 7369–7381.

Baker, R. S. J. d. Data Mining for Education. In *International Encyclopedia of Education,* 3rd ed.; McGaw, B., Peterson, P. Baker, E., Eds.; Elsevier: Oxford, 2011.

Baker, R. S. J. d.; Yacef, K. The State of Educational Data Mining in 2009: A Review and Future Visions. *J. Educ. Data Min.* **2009,** *1*(1), 3–17.

Bienkowski, M.; Feng, M.; Means, B. *Enhancing Teaching and Learning Through Educational Data Mining and Learning Analytics: An Issue Brief.* Department of Education, Office of Educational Technology: Washington, D. C., U.S., 2012.

Borne, D. K. Top 10 Big Data Challenges—A Serious Look at 10 Big Data V's. https://www.mapr.com/blog/top-10-big-data-challenges-%E2%80%93-serious-look-10-big-data-v%E2%80%99 s (accessed Apr 11, 2014).

Boyd, D.; Crawford, K. Critical Questions for Big Data. *Inf. Commun. Soc.* **2012,** *15*(5), 662–679. DOI: 10.1080/1369118X.2012.678878.

Gewirtz, D. Volume, Velocity, and Variety: Understanding the three V's of big data. http://www.zdnet.com/article/volume-velocity-and-variety-understanding-the-three-vs-of-big-data/ (accessed Apr 20, 2016).

Goes, P. B. Big Data and IS Research. *MIQ Q.* **2014,** *38,* 3.

Harper, E. M.; Parkerson, S. Powering Big Data for Nursing Through Partnership. *Nurs. Adm. Q.* **2015,** *39*(4), 319–324.

Hilbert, M. Big Data for Development: From Information- to Knowledge Societies, 2013. http://ssrn.com/abstract=2205145 or http://dx.doi.org/10.2139/ssrn.2205145 (accessed Sept 10, 2016).

Holland, P. *Characteristics of Big Data—Part One*. Retrieved from Data Intensity. (accessed May 26, 2015).

Kim, Y. H.; Ahn, J. H. A Study on the Application of Big Data to the Korean College Education System. *Procedia Computer Science* **2016,** *91,* 855–861.

Leu, F. Y.; Liu, C. Y.; Liu, J. C.; Jiang, F. C.; Susanto, H. S-PMIPv6: An Intra-LMA Model for IPv6 Mobility. *J. Network Computer Applications* **2015,** *58,* 180–191.

Leu, F. Y.; Ko, C. Y.; Lin, Y. C.; Susanto, H.; Yu, H. C. (2017). Fall Detection and Motion Classification by Using Decision Tree on Mobile Phone. In *Smart Sensors Networks;* Xhafa, F., Leu, F.-Y., Hung, L.-L., Eds., Academic Press, 2017, pp 205–237.

Linoff, G. S.; Berry, M. J. *Data Mining Techniques: For Marketing, Sales, and Customer Relationship Management;* Wiley Computer Publishing, 2011.

Liu, J. C.; Leu, F. Y.; Lin, G. L.; Susanto, H. An MFCC-Based Text-Independent Speaker Identification System for Access Control. *Concurrency Computation: Practice Experience* **2018,** *30*(2), e4255.

Manyika, J. *Big Data: The Next Frontier for Innovation, Competition, and Productivity;* Mickinsey Global Institute, 2011.

Mayer-Schonberger, V.; Cukier, K. *Big Data: A Revolution That Will Transform How We Live, Work and Think;* John Murray Publishers: U.K., 2013.

Picciano, A. G. The Evolution of Bid Data and Learning Analytics in American Higher Education. *J. Asynchronous. Learn. Networks* **2012,** *16*(3), 9–20.

Reyes, J. The Skinny on Big Data in Education: Learning Analytics Simplified. *TechTrends* **2015,** *59*(2), 75–80.

Rijmenam, M. v. Why The 3V's Are Not Sufficient To Describe Big Data. https://datafloq.com/read/3vs-sufficient-describe-big-data/166 (accessed Aug 7, 2013).

Romero, C.; Ventura, S. Educational Data Mining: A Review of the State of the Art. *IEE Trans. Syst. Man Cybern. C: Appl. Rev.* **2010,** 40(6), 601–625.

Rossi, B. *Top 4 Ways to Apply Big Data in Higher Education.* http://www.information-age.com/it-management/strategy-and-innovation/123460114/top-4-ways-apply-big-data-higher-education (accessed Jun 7, 2018).

Schmarzo, B. *What Universities Can Learn from Big Data – Higher Education Analytics.* https://infocus.emc.com/william_schmarzo/what-universities-can-learn-from-big-data-higher-education-analytics/ (accessed July 2, 2014).

Soubra, D. The 3 Vs that define Big Data. http://www.datasciencecentral.com/forum/topics/the-3vs-that-define-big-data (accessed July 5, 2012).

Strausser, J. Examining the application of big data analytics to increase university retention and promote student success, 2015.

Susanto, H. Managing the Role of IT and IS for Suppoting Business Process Reengineering. *J. Systems Information Technol.* **2016.**

Susanto, H. Cheminformatics—The Promising Future: Managing Change Of Approach Through ICT Emerging Technology. In *Applied Chemistry and Chemical Engineering, Volume 2: Principles, Methodology, and Evaluation Methods*; Haghi, A. K., Pogliani, L., Balkose, D., Mukbaniani, O. V., Mercader, A. G., Eds.; Apple Academic Press, 2017a,pp 313–332.

Susanto, H. Biochemistry Apps as Enabler of Compound and DNA Computational: Next-Generation Computing Technology. In *Applied Chemistry and Chemical Engineering, Volume 4: Experimental Techniques and Methodical Developments;* Haghi, A. K., Pogliani, L., Castro, E. A., Balköse, D., Mukbaniani, O. V., Chia C. H., Eds.; 2017b, Apple Academic Press, pp 181–202.

Susanto, H. Electronic Health System: Sensors Emerging and Intelligent Technology Approach. In *Smart Sensors Networks;* Xhafa, F., Leu, F.-Y., Hung, L.-L., Eds., Academic Press, 2017c, pp. 189–203.

Susanto, H. Smart Mobile Device Emerging Technologies: an Enabler to Health Monitoring System. *Kalman Filtering Techniques for Radar Tracking* **2018,** 241.

Susanto, H.; Almunawar, M. N. Managing Compliance with an Information Security Management Standard. In *Encyclopedia of Information Science and Technology,* 3rd ed.; Khosrow-Pour, M., Ed.; IGI Global, 2015, pp 1452–1463.

Susanto, H.; Almunawar, M. N. Security and Privacy Issues in Cloud-Based E-Government. In *Cloud Computing Technologies for Connected Government;* Mahmood, Z., Ed.; IGI Global, 2016, pp 292–321.

Susanto, H.; Almunawar, M. N. *Information Security Management Systems: A Novel Framework and Software as a Tool for Compliance with Information Security Standard.* CRC Press, 2018.

Susanto, H.; Chen, C. K. Information and Communication Emerging Technology: Making Sense of Healthcare Innovation. In *Internet of Things and Big Data Technologies for Next Generation Healthcare;* Bhatt, C., Dey, N., Ashour, A. S., Eds.; Cham: Springer, 2017, pp 229–250.

Susanto, H.; Chen, C. K. Macromolecules Visualization through bioinformatics: an Emerging Tool of Informatics. *Appl. Phys. Chem. Multidiscip. Approaches* **2018a,** 383.

Susanto, H.; Chen, C. K. Informatics Approach and Its Impact for Bioscience: Making Sense of Innovation. *Appl. Phys. Chem. Multidiscip. Approaches* **2018b,** 407.

Susanto, H.; Almunawar, M. N.; Leu, F. Y.; Chen, C. K. Android vs iOS or Others? SMD-OS Security Issues: Generation Y Perception. *Int. J. Technol. Diffusion* **2016a,** *7*(2), 1–18.

Susanto, H.; Kang, C.; Leu, F. Revealing the Role of ICT for Business Core Redesign, 2016b.

Susanto, H.; Chen, C. K.; Almunawar, M. N. Revealing Big Data Emerging Technology as Enabler of LMS Technologies Transferability. In *Internet of Things and Big Data Analytics Toward Next-Generation Intelligence;* Dey, N., Hassanien, A. E., Bhatt, C., Ashour, A. S., Satapathy, S. C., Eds.; Cham: Springer, 2018, pp 123–145.

Vincent, K. O. Business Intelligence and Big Data Analytics for Higher Education: Cases from UK Higher Education Institutions. *Inf. Eng. Express* **2016,** *2*(1), 65–75.

CHAPTER 3

REVEALING THE BIG DATA REVOLUTION AS A DRIVER OF SCIENCE ACHIEVEMENT IN TODAY'S GLOBALIZATION

HERU SUSANTO[1,2,*] and FANG-YIE LEU[2]

[1]*The Indonesian Institute of Sciences, Indonesia, E-mail: heru.susanto@lipi.go.id*

[2]*Tunghai University, Taichung, Taiwan, E-mail: leufy@thu.edu.tw*

[*]*Corresponding author. E-mail: heru.susanto@lipi.go.id*

ABSTRACT

Big data, most often considered as "crude oil," has a large volume of data that comprises raw data that has not been processed. On the other hand, data science means "data mining." It is the extraction of big data using means of data analysis. Understanding big data to support data science can be referred to the extraction of big data with the help of critical analysis, and this extraction results in the 4Vs of big data roles. This report focuses on understanding big data to support data science, which started with the research and analysis on the roles of big data and data sciences through various literature reviews. Followed by critical analysis, the result and discussion are conducted on the benefits and issues of the research studies. In conclusion, recommendations are given to solve the identified problems. However, there are limitations in the research as there were only a few studies conducted on data science. As concluded, without big data, data science can still function but it will neither give out the right nor correct or complete information. It is understandable that data science has the ability to create potential knowledge from the data extraction and also brings more benefits, too, but the rising of several issues should also be tackled.

3.1 INTRODUCTION

Big data studies have begun since the year 2001, the studies have shown that the amount of data continuously increases as the technology and time passes. Big data, most often considered as "crude oil," has a large volume of data that comprises raw data that has not been processed (Provost and Fawcett, 2013).

The advancement of big data has reached the business practice has been progressing the technology of information system, as well as data science (Liu et al., 2018; Susanto, 2017a; 2017b; Leu et al., 2015; 2017). Using big data is an advantage to any business, be it a start-up or already existing business. Previously known as data warehouse, big data evolved to assist business in many ways such as marketing and management (Susanto, 2016; Susanto et al., 2016b; Almunawar et al., 2013a; 2013b).

Characteristics of big data is make up of 4Vs; veracity, volume, velocity, and variety. Big data's 4Vs play a major role in modern information system practices. The increasing usage of Internet and the volume of big data provide an opportunity for businesses like Starbucks as well as the current trend "Internet of things" in which cars, smartphones, and television are used as a medium for marketing and payment itself. Big data also originated from the usage of information system. For example, using Facebook, it records the cache and cookies of the Facebook user and stores this in the user's hardware files. This cookie and cache are big data that stores the information of the user and is used to direct advertising according to the criteria of the user which mostly shows the user's preferences (Susanto and Chen, 2018a; 2018b).

Based on the study of "big data in health informatics," big data is the database for patients' records, monitoring patient's condition, information on various other alternatives of medical practices, and checking if there is any error of data (Susanto, 2017c; 2018; Susanto and Chen,2017).

On the other hand, data science means "data mining." It is the extraction of big data using means of data analysis. Based on a research on data science, it is the combination of statistical analysis, programming, and domain expertise (Conway, 2013). Supporting data science with big data refers to the extraction of big data with the help of critical analysis and this extraction results in the 4 Vs of big data roles.

According to Carbone et al. (2016), big data and data science work hand-in-hand as data science has the potential to extract valuable and meaningful data from various sources to be used for knowledge. In this report, we will be focusing on the understanding of big data to support data science. The beginning part of the report will be the research and analysis on the roles of big data and data sciences through various literature reviews. Later, critical analysis is conducted on the benefits and issues of the research studies in the result and discussion. In conclusion, recommendations are given to solve the identified problems.

3.2 LITERATURE REVIEW

3.2.1 BIG DATA

Big data as the name itself refers to big in sizes in the form of unprocessed data. Over the past few years, as Internet evolves and changes, the volume of data collected from all over the world has increased rapidly due to the advancement of technology and the usage of Internet users over time. Not only big data can be found online, but it can also be collected from offline application through collection of applications on Android devices or internal data collection that is unreleased to the public.

From Carbone et al.'s (2016) analysis from the above definition for big data, it only fit one of the characteristics of big data, which is volume. Collection of bigger data can lead to the results becoming more illustrative, but more storage space is needed to store that large capacity and more computational inputs will be required for data analytics.

The big data is mined into four types of categories such as veracity, velocity, volume, and variety. From these categories, traditional methods of data mining cannot be used as it cannot ensure the safety of the data results. Further explanation on the 4 Vs is given in the later part of data science.

3.2.2 DATA SCIENCE

The new technology that can mine big data is called data science. It can be a program or another kind of data analyzing software. McAfee and

Brynjolfsson (2012), cited by Carbone et al. (2016), have proven that interest toward big data is not based on the procedure and managing the huge data capacity. Preferably, it is the potential outcome of knowledge that releases from the integration of several data sources. The rise of big data usage increases the awareness to protect privacy of users, and to avoid any unnecessary malicious software.

Each of big data categories carries its own advantages. In volume case, the amount of big data gathered, and contained is analyzed for the records of past accumulation, for example, usage of social media itself. For example, from the usage of Facebook, over the years this social media has evolved from just getting to know people all over the world into a business medium. This is because of the influence on big data. The cache and cookies collected from each user's data provided what kind of advertising, friends, or what type field the user wants to see every day.

As for velocity, again, taking Facebook as an example, users use social media most of the time to update on their social status or checking on their friends' status, even for Instagram that has a mechanism that connects their posts to Facebook, too. This accumulated data is generated every second of the day, fast and efficient. Users also respond to any changes and new development of these social media which further evolves the business strategy.

In the matter of veracity, it is the concerning error of the data or any complications that can be gathered from the data itself. Finally, the variety of big data which stores the variation of data sources, for example, one user can accumulate "what news needs to be seen" or "what kind of people recommended to add," these data comes into one whole big chunk of data.

For a business, this big data is a tool for strategical redesign because of the variation of the data. Take Facebook as a sample, it generates its profit from advertising with a selected audience type of advertising. For example, Facebook requires the user to choose what kind of advertising they want to see; what kind of friends they have; what kind of advertising pages they had liked; and then these data are accumulated through the feedback of users which is done from time to time by Facebook to each and every customers.

The mining of big data from "online" can be easy than an "offline" which is a challenge because the user is not using the application online or the user is a concern on security and privacy issues making all mined data cannot be a complete set (Susanto and Almunawar, 2018; 2016; 2015).

Big data is not only used in businesses but also in medical as well as a latest trend "Internet of things." One of the mechanisms that stores big data inside a small chip is called radio-frequency identification (RFID) which installed in items, for example, Walmart that is using this technology to know what kind of items their customers often buy, what range of age, or the time when their customers shops the most. Other forms can be "Internet of things" which is practiced in Japan; customers can pay for their drinks from vending machine by scanning their phone this is called as "digital wallet."

The variation of big data in many fields has increased the issue of security and privacy despite the data is collected from anonymous users (Bertino, 2016). Thus, various types of data science are developed to overcome the issue.

Data science was initially known as statistics or data mining, but as big data evolves, a few scientists have adopted new ways of encrypting data. In business field, big data originating from social media, it needs a large storage and processor, for example the Microblog processing toolkit (MPT) (Jie Wang, 2015). In medical field, it is "health informatics" a combination of computer science and information science software for the realm of health care.

The need of big data is mutually exclusive with the need of data science. Big data is present almost everywhere, and it keeps on evolving in which data science plays a big role in decoding it.

Gathering big data nowadays is easier compared to older days because of advancement in technology especially through smartphones. A smartphone operates through Android system which is an open source system and the user can modify the data. Third-party applications like Messenger or Instagram accumulates big data in high volume and how these applications gather the information from the users is by the system that operates it itself.

Despite the issue on security and privacy, big data still plays an important role in information system strategies. The dependence of using big data in business is very crucial to gather customer information, in order to deliver the right information to customers as well as to develop business to customer relationship. For example, online business or e-business deals with the supplier directly to customer with eliminating "salesperson" intermediate and re-intermediating "bank" as medium of transaction of online business. The bank will also receive big data from customers who often spend their credit or debit card from online purchase so they can pinpoint which type of business requires online transaction and charges commission from it.

Understanding how big data works is by understanding how to use data science. To survive the era of modern technologies, to speed up business practices, creating more business and job opportunities, big data must be present in order to accommodate the activities and performance. In order to achieve success in a business, big data supported by data science into making a data-driven decision making is needed as it provides information and engineering at big scale (Provost and Fawcett, 2013).

3.2.3 MACHINE LEARNING (ML)

Machine learning (ML) is "a sub-field of artificial intelligence (AI)" focuses on using computational systems to analyze data in finding out ways to perform a desired task automatically (Ali et al., 2016). Ali et al. (2016) refers ML as a system that is used to create decision-making, making prediction, or future trends, and it is the key that applies technology to various fieldwork like healthcare, science, engineering, business, and finance in introducing data mining and big data techniques.

According to Ali et al. (2016), ML tasks can be categorized into supervised, unsupervised, reinforcement, deep, and numeric learning. The task for each learning techniques varies according to the data that will be researched and analyzed to find specific outcome.

Supervised learning technique focus on the generalized outcome in which the outcome can be continuous to assume any value or descriptive which will describe the data. Meanwhile, unsupervised learning task uses clustering which is the learning task which only collects data that perceive similar characteristic without the needs of labeling (Ali et al., 2016).

In reinforcement learning, using machine language gives either positive or negative outcome out of the data that has been used. Deep learning uses machine language to solve too complex problems that general minds cannot understand (Bennet and Bennet, 2008). Based from Witten and Frank (2005), cited by Ali et al. (2016), numeric learning by using machine language can perform task on the data to give numeric output.

3.2.4 DATA ANALYTICS

Analytics refers to the "techniques used to analyze and acquire intelligence from big data" (Gandomi and Haider, 2014). As big data comprised large volume of raw data, analytics techniques help to unleash the potential of the data and transform into useful data. Currently, many different types of software are being designed to analyze data. Large organizations namely SAS Institute and SPSS Inc., in particular, were creative inventors in menu-based statistical analysis software. Their inventions turned out to be successful as the software inventions can allow unprofessional analysts to perform the complex statistical analysis (Agrawal, 2014).

IBM conducted a survey about innovation to more than 1000 business leaders and their findings conclude that big data and analytics have played important roles in organizations that seek innovation (Marshall et al., 2015). Marshall et al. (2015) derive that an organization which uses big data and analytics can outperform other competitors in terms of income growth and effective daily operation of the organization. Moreover, due to the process of analyzing and extracting valuable data, it can improve the efficiency of the organization, as well as configure the operation by being innovative to outperform their competitors in terms of product or services.

Analytics is not a new technique and it has been around since 1950s. Before 2003, data analytics only works with limited data that only come from the internal findings and the data is stored in the data warehouse. The outcome of these analytics was descriptive. With the emergence of big data in 2003, technology firms such as Google and Yahoo use big data for internal analysis and using customer focus strategies (Larson and Chang, 2016). Large organizations are focusing on creating a data-driven economy by analyzing the big data.

Big data analytics refer to a part of the overall process of "insight extraction" from big data. Further given is a list of some techniques of big data analytics that are being used for structured and unstructured data. The list of techniques of big data analytics consists of text, audio, video, social media and predictive analytics (Gandomi and Haider, 2014).

Text analytics refers to the technique of extracting information from written based data. It can also be known as text mining which these techniques involve statistical analysis, computational linguistics, and ML. The type of information that organization can use to extract is from website, social media, survey, documents, and news. Text analytics enables organizations to summarize all the textual information to provide a proof for decision making.

Audio analytics used to extract unstructured data which in the forms of audio and analyze it. Audio analytics can be defined as speech analytics as well, by which the terms can be used interchangeably. For businesses, audio analytics information currently only available from customer care services, a conversation between workers and customers. Audio analytics gains inside information which organization can analyze the findings and convert it into recommendations to improve the businesses strategies and next stages of business processes (Gandomi and Haider, 2014).

Video analytics or video content analysis (VCA), is a technique that extracts the information by monitoring and analyzing from videos. Panigrahi et al. (2010), cited in Gandomi and Haider (2014), commented that "data analytics is still in its infancy compared to other types of data mining." Monitoring videos are no longer only from prerecorded but, with the latest technology development like closed circuit television (CCTV) camera, data analysts can monitor and analyze the currently ongoing situation in real time.

For social media, analytics refers to analyzing the social media to extract information. It can be in the form of structured or unstructured data. Social media has the capabilities to transmit information from one online platform to another which enables the users to create and share their information easily. Social media analytics covers a broad range of social media platforms; social networks, blogs, social news or wikis, to extract and analyzing the information. Social media analytics is still considered as a new technique as social media only started in early 2000 with the introduction of Web 2.0 (O'Reilly, 2005).

Predictive analytics from the name itself refers to a variety of techniques to forecast the future with the use of historical and current data. The objective of predictive analytics is to provide or give a competitive advantage by assuming the possible events occurring in the future (Ali et al., 2016). Mostly the information used for predictive analytics is primarily based on statistical methods from big data.

SAS was identify the importance of big data analytics as it helps organizations harness their data and use to identify opportunities. The Davenport research clarifies how businesses used their data, his findings that big data analytics helps to reduce cost, an improvement in the decision making in a faster way, new products and services development.

For cost reduction, using existing technologies for big data like Hadoop and cloud-based analytics can help to store a large volume of data at a price that is cheaper than buying the personal database. By using Hadoop for the speed and in-memory analytics, with the analysis of new information, organizations or businesses can make a decision at once from the data that have been collected. With big data analytics, gathering of information can help to learn what customer wants, which businesses can take advantage of this ability to create new products or services that meet the demand of the.

Schutt and O'Neil (2013), cited by Larson and Chang (2016), clarifies that by applying data science, analysis can be done with statistical measure and using machine language algorithms to "produce data products or models that perform descriptive, predictive, or prescriptive analysis." These changes on how the traditional data analysts perform to the modernize data analyst in terms of the outcome analysis. Furthermore, the speed of analyzing data becomes faster with the use of integrated technology.

3.3 RESULTS AND DISCUSSION

The result and discussion section emphasizes on two parts. The first part focuses on the implication of understanding big data to support data science by elaborating on the benefits and followed by the drawbacks toward using of big data from the process ML, data mining, and data analytics as a whole.

The advantages of understanding big data to support data science:

1. New level of knowledge and research outcome
   Big data have a lot of potential data that can be converted into useful and reliable source by data science. For that, data that have been extracted by data science give a whole new level of knowledge and research outcome that needs to be explored. Collecting contextual data is now easily available and inexpensive, these type of information, contextual information creates a potential for new knowledge in business, customers or other factors that can be useful to the organization.

   Contextual information "provides important clues to understand why different outcomes occur in different settings." Contextual information can be collected in a spatiotemporal setting (e.g., to customers going in and out of the shopping complex) or a sequence of settings over time (e.g., survey is conducted to find out the customers' preference of shopping area on certain days or time by using the location of their mobile device). This research can give a more beneficial outcome that discovers the consumer behaviors that was never been looked at and known before. Furthermore, information collected can go beyond to individual level by finding out the frequent time the consumer visiting e-commerce website and study consumers' rapidly purchase to time and products or responses to tweets or giving feedback in the social media like Twitter or Facebook on responses about the services that have been provided to them. Instantaneously, consumer-decision responses are triggered by the data collection (Chang et al., 2013; Almunawar et al., 2018a; 2018b; 2015; Susantp et al., 2016a).

2. Instant data visualization
From data science, especially data mining, it is easier to analyze from the collection of big data due to the ability of prediction. In the context of education, as an example, if a school administration wants to identify a student that seeks help in terms of study or the school wants to decrease student retention, data mining has the ability to predict the outcome from the collected data and instantly can be visualized by charts and graphs. This saves the time and improves the efficiency in delivering education to the students (Ranjan and Malik, 2007). Given that data have been collected and stored, data science can process the data and produce an outcome in the ways that the users wanted. For that there is the need to use the decision-making tools to provide the instant data visualization (Susanto et al., 2018).
3. Decision making tools
As a start scientist developed various tools that can help make the decision easier from the big data extraction. By analyzing literature review, the decision-making tools can be used in various fields of work; education, health, businesses, or governments, in which the tools have the effects of extracting, analyzing, explaining, and visualizing big data. Here are some of the results from the tools that can be in the form of numerical or statistics for mathematical techniques, complex data solving using AI for data analysts, and also, in the form of graphs and pictures for

visualization technique. It brings more benefits as the tools are user friendly which allows users who lacks in a certain field of knowledge and are not experts in using the tools to get a different form of information. In return, the information obtained is able to become a problem solver.

Next, the disadvantages are discussed below:

1. Privacy and security issues
   Using big data and gathering it for marketing and sales strategies may possess a great potential but it is also a drawback. Collection of data without the consent of the user that includes the privacy of user, what kind of daily life the user goes through, what kind of business transaction a user does, or where does the user live, it is already a breach of privacy. In terms of security issues, there can be misleading use of the information gathered by which the data is being transferred or shared with another organization without the consent of the people, too. There can also be losses of data or data leakage occurring if the information is not being handled properly by the right people (Chang et al., 2013).
2. Complicated in collecting data and extraction
   Big data collection and extraction may not be easy for some as the volume is too large to handle. Advanced analytical program or software might be required to gather the information needed and in finding the reliable information that can be used for the extraction of information (Jifa and Lingling, 2014).
3. Money involvement to get reliable data
   Use of big data can be helpful. However, getting the information that is useful to the research may be costly. Dealing with data can be related to money involvement. This means that researchers have to pay a lot of money to get the useful and reliable data in order to collect the real data which sometimes involving the privacy of ordinary people (Jifa and Lingling, 2014).
4. Storage in the cloud
   Storing in the cloud can be quite useful for the access of the materials later on when it is needed and saved up some space in hardware devices. However, there is a disadvantage of these services. Once the data is uploaded to the server online like cloud

computing, there is a chance that the data information stored in the cloud cannot be deleted. As the data is being stored somewhere the users are unable to know. Even if users manually delete their information, other users might have already downloaded the data that was uploaded, and making multiple copies of the data.

3.4 LIMITATIONS

The study of big data and data science has started way back 20th century, as the Internet evolves. There is no exact definition of data science as the terms can change from different researchers' point of view. However, there are very few studies of data science have been conducted (Mellody, 2014), and there is no specific data science that can properly decode all big data.

Big amount of data does not ensure the completion or meeting the target research. There could be misleading or unnecessary information gathered for example, "personal opinion" which does not concern business or marketing suggestions.

3.5 CONCLUSION

As a conclusion, big data is a subset of data science; it supports the functions of data science. Without big data, data science can still function but it will not give out the right or correct information or complete information. The importance of big data in the veracity, velocity, value, and variety elements, drives data science to produce data-driven decision-making, creating more information for customer relationship and speeding up business performance on a wide scale.

The evolution of big data has created a competitive advantage for all types of business as well as in health care and services. Behind the unprocessed raw data, data science reengineered big data into a more valuable piece of information by further analyzing and mining it into a tool for information strategies.

Last, it is understandable that data science has the ability to create potential knowledge from the data extraction and also bringing more benefits, too, but the rising of several issues should also be tackled.

3.6 RECOMMENDATION

Big data is now a major icon in business and a tool for competitive advantage but cannot be underestimated. To combat the security and privacy issues with the increasing volume of data, security enhancement must also be tightened and updated with correlation to its usage. Personal data that is collected by organizations need to be anonymous so that no personal identifying information is publicly available.

It is just not business practices alone using big data; it can also be introduced as an education scheme to create awareness among the young generations to make use of big data in a more meaningful way. By creating tools that can be beneficial toward the studies in order to increase the knowledge of young generation. Relatively, having a filter technology to make selective data becomes available as the volume of data increases, making extraction might be time consuming for the students.

The vast variety of big data from data mining can be expanded to communication field or using big data as a tool for communication. So far, the programs are only available for the use of mining and translating information as these programs are able to make big data to be more structured and data being organized properly, there is a possible invention in data mining to expand into communication field.

KEYWORDS

- **data analytics**
- **data science**
- **internet of things**
- **machine learning**
- **social media**

REFERENCES

Agrawal, D. Analytics Based Decision Making. *J. Indian Bus. Res.* **2014,** *6*(4) 332–340. http://dx.doi.org/10.1108/JIBR-09-2014-0062

CHAPTER 4

TOWARD BIG DATA'S IMPACT ON THE LEARNING PROCESS: GENERATION Y AND Z PERSPECTIVES

HERU SUSANTO

Indonesian Institute of Sciences, Jakarta, Indonesia,
E-mail: heru.susanto@lipi.go.id

Tunghai University, Taichung, Taiwan

ABSTRACT

In the information era, enormous amounts of data have become available on hand to decision makers. "Big data" as a term has been among the biggest trends over the last few years, leading to an upsurge of inquiry, as well as industry and political science diligence. Big data refers to datasets that are not only big but also high in variety and velocity, which makes them difficult to handle using traditional tools and techniques. Data is deemed a powerful raw material that can impact multidisciplinary research effort as well as government and business performance. The goal of this discussion paper is to reveal the "big data" for Generations Y (Gen Y) and Z (Gen Z) for learning process brought forth by the big data cause. The authors from different journals will bring together diverse linear perspectives and additional sentiment, coming from different background environment, geographical positions, and believes with different nucleus-like research expertise and different affiliations, alliances, and workplace experiences. This paper aims to analyze some of the different analytics methods and tools which can be applied to big data, as well as the different uses of "big data" in both Gens Y and Z.

4.1 INTRODUCTION

4.1.1 *BIG DATA*

The term "big data" is sometimes considered as an industry or a technology by many but it is neither of these; it is a term that explains data that cannot be processed or analyzed using normal traditional techniques in a cost-effective and timely means. Moreover, it describes a range of data, data types, and tools to address the quickly increasing amount of information that organizations across the world are managing. The total of data obtained, kept, and processed by this varied range of organizations has grown exponentially. This has been driven, in part, by an explosion in the amount of data obtained from web-based businesses, social media, and sensors from devices.

Numerous ways are used by organizations to create value through big data. The data collected can be used to come up with a better understanding of customers and to create products and services for narrowly distinct sectors. Big organizations can use these data to monitor the performance of key functions, classifying issues contributing to observed alterations, and highlighting required corrective actions or new ways to enhance the systems. Some use data to forecast behavior or predict events, and as a result, take suitable acts. Data can contribute in helping to meet regulatory agreement or legal discovery requirements. Finally, organizations can use data as the building blocks for new products and services found throughout the industries.

While the term "big data" is fairly new, the act of collecting and keeping huge amounts of information for eventual study is ages old. The concept gained momentum in the early 2000s when industry analyst Doug Laney voiced the now-mainstream meaning of big data as the three Vs and two added scopes.

4.1.1.1 VOLUME

Volume refers to massive information sets that are on the order of scales larger than data managed in the old-style databases. Though the general scale of data being collected and stored is certainly striking, the real issue is the amount of data handled by different organizations.

For example, Facebook has more than a billion active users with 150 billion friend connections. Every bit of new content, ranging from news feeds, messages, events, photos, and ads, is kept and tracked along with the enormous amount of data contained in websites. More than 500 TB of new data are loaded into the company's records every day with the largest Hadoop cluster with the capability of keeping more than 100 PB. The need to keep and process huge amount of data is not limited to commercial concerns.

Organizations gather data from a diversity of sources, including business transactions, social media, and information from sensor or machine-to-machine data. In the past, storing it would have been a problem—but new technologies (such as Hadoop) have relieved the problem.

Another example, the Large Hadron Collider generates ~15 PB of data per year. Moreover, the planned Large Synoptic Survey Telescope will produce around 20 TB of data per night, resulting in 60 PB of raw data and a catalog database of 15 PB over 10 years of business operations. The whole volume of data after processing will be more than a several hundred PB.

4.1.1.2 VELOCITY

There are a couple features of the need for speed. The first focuses on the ability to process data as they arrive. While some data are produced periodically, others such as machine data are delivered in a constant stream.

Taking the Large Hadron Collider as an example again, the 150 million sensors in the facility deliver data 40 million times per second. The second feature refers to how quickly data need to be processed. While processing historical data for business intelligence reporting or more in-depth analysis might need to be completed within minutes or hours, other tasks are more time sensitive. Certain types of transactions such as processing a trade or placing a targeted ad require the ability to process data in milliseconds.

Data streams in at an extraordinary speed must be handled within a timely way. Radio Frequency Identification tags, sensors, and smart metering are driving the need to deal with streams of data in near-real time.

4.1.1.3 VARIETY

The increase in volume has been accompanied by an increase in the complexity of data that organize store and process. Recently, attention was focused on structured data, that is, data that are precisely arranged based on a predefined formal plan (e.g., relational database). However, most data do not fit this description.

Data derives in all types of formats—from structured, numeric data from old-style databases to unstructured text documents, e-mail, video, audio, stock ticker data, and financial dealings.

The term semi-structured data indicates a mixture of structured as well as unstructured components. This may include, for example, XML and other markup languages.

4.1.1.4 VARIABILITY

In addition to the increasing speeds and diversities of data, data flows can be highly varying with periodic peaks. Therefore these daily, seasonal, or event-triggered peak data loads can be stimulating to handle. Even more so with unstructured data.

4.1.1.5 COMPLEXITY

Data coming from multiple sources are difficult to link, match, cleanse, and transform data across systems. However, linking and correlating relationships, hierarchies, and multiple data connections are essential or the data collected can quickly spiral out of control.

4.1.2 IMPORTANCE OF BIG DATA

The significance of big data not only involves how much data a company has, but what they do with the data gathered. Taking data from any source and analyzing it will enable a company to:

1. Save cost
2. Reduce time wastage

3. Develop new products and services as well as optimize offerings
4. Make smart and wise decisions

When combining big data with high-powered analytics, that company can accomplish business-related tasks such as

- Determining original causes of failures, problems, issues, and defects in near-real time
- Generating coupons and promotions at the point of sale based on the customer's buying behaviors
- Recalculating the company's whole risk portfolios in minutes
- Spotting fraudulent behavior before it affects the rest of the organization

4.1.3 SOURCES OF BIG DATA

- Archives: Archives are scanned documents, statements, insurance forms, medical record and customer correspondence, paper archives, and print stream files that contain original systems of record between organizations and their customers.
- Docs: XLS, PDF, CSV, e-mail, Word, PPT, HTML, HTML 5, plain text, XML, JSON, and so forth.
- Media: Images, videos, audio, flash, live streams, podcasts, and so forth.
- Data Storage: SQL, NoSQL, Hadoop, doc repository, file systems, and so forth.
- Business apps: Project management, marketing automation, productivity, CRM, ERP content management systems, HR, storage, talent management, procurement, expense management, Google Docs, intranets, portals, and so forth.
- Public web: Government, weather, competitive, traffic, regulator, compliance, healthcare services, economic, census, public finance, stock, OSINT, the World Bank, SEC/Edgar, Wikipedia, IMDb, and other Web services.
- Social media: Twitter, LinkedIn, Facebook, Tumblr, Blog, SlideShare, YouTube, Google+, Instagram, Flickr, Pinterest, Vimeo, Wordpress, IM, RSS, Review, Chatter, Jive, Yammer, and so forth.

- Machine log data: Event logs, server data, application logs, business process logs, audit logs, call detail records, mobile location, mobile app usage, clickstream data, and so forth.
- Sensor data: Medical devices, smart electric meters, car sensors, road cameras, satellites, traffic recording devices, processors found within vehicles, video games, cable boxes or household appliances, assembly lines, office buildings, cell towers and jet engines, air-conditioning units, refrigerators, trucks, farm machinery, and so forth.

4.1.4 USES OF BIG DATA IN DIFFERENT INDUSTRIES

Organizations are using big data to create new products and generate insights into a wide range of phenomena. Applications are widespread, including fraud detection, customer sentiment analysis, ad personalization, stock trading, drug discovery, healthcare delivery, energy efficiency, and management of computer and telecommunication networks.

There are five broad ways in which organizations can use big data to create value

1. Organizations can use data to develop a better understanding of their customers and tailor product and services for narrowly defined segments.
2. Organizations can use data to monitor the performance of key functions, identifying factors contributing to observed variances, and pointing to needed remedial actions or ways to optimize systems.
3. Organizations can use data to predict behavior or forecast events, and as a result, take appropriate action.
4. The organization can use data to meet regulatory compliance or legal discovery requirements.
5. The organization can use data as the building blocks for new products and services.

The following are other examples of uses of "big data" in different industries

- Financial services: fraud detection and analysis, credit risk scoring and analysis fraud detection and analysis, credit risk scoring and analysis, risk analysis and mitigation, automated trading algorithms, compliance and regulatory reporting, legal discovery, customer sentiment analysis, targeting product offerings

- Marketing and advertising: customer sentiment analysis, campaign analysis, trading/pricing of ads, Personalized web content/e-mails, ad targeting/serving
- E-commerce/retail trade: click stream analysis, customer sentiment analysis, analysis of buying behavior, call center/log analysis, point of sale transaction analysis, development and application of pricing models, personal web content delivery, ad targeting/serving, inventory management
- Social media: ad targeting/serving, customized content and promotion, location-based services
- Media and entertainment: customer sentiment analysis, content streaming, search and recommendation optimization, customized content and promotions, ad targeting/serving
- Telecommunication: customer sentiment analysis, analysis of buying behavior, analysis of usage patterns, call center/log analysis, location-based service, network analysis and optimization, predictive maintenance
- Manufacturing: process control, capacity utilization and forecasting, supply chain analysis and management, predictive maintenance, analysis of warranty claims
- Transportation: location tracking, capacity utilization and forecasting, development and application of pricing models, fuel consumption analysis, predictive maintenance
- Energy and utilities: smart meter analytics, compliance audits, real-time demand forecast and pricing, network analysis and optimization, predictive maintenance
- Healthcare: clinical trials in silico, comparative effectiveness, social media analysis to detect disease or treatment patterns, capacity utilization and forecasting, patient monitoring, personalized medicine, billing compliance.
- Life sciences: genomic sequencing, drug discovery, drug surveillance/monitoring
- Government: fraud detection and analysis, threat analysis, analysis of crime patterns, weather forecasting, cybersecurity

4.1.5 BIG DATA BUSINESS SEGMENTS

4.1.5.1 BIG DATA APPLICATIONS

Some of the big data applications are healthcare, telecom, financial services, transportation, life sciences, social media, e-commerce/retail, energy/utilities, government, media/entertainment, marketing/advertising, and manufacturing.

4.1.5.2 BIG DATA TECHNOLOGY PLATFORM

Some of the examples of big data technology platform are as follows:

1. Analytic tools: business intelligence, statistical analysis, data visualization
2. Data management tools: next generation data warehouse, HDFS/MapReduce, data integrated tools, NoSQL/news SQL databases
3. Hardware: storage, servers, networks
4. Service providers: consulting/training, system integration, cloud service provider

4.1.5.3 DATA SOURCES

Some of the examples of data sources are documents, images, video, real-time, transactional, sensors, and Internet of things (IoT).

4.1.6 BIG DATA ANALYTICS

It examines large amounts of data to uncover hidden patterns, correlations, and other insights. With today's technology, it is possible to analyze data and get responses from it almost instantly—an effort that is slower and less well-organized with more old-style business intelligence results.

Most establishments now know that if they obtain all the data that streams into their businesses, they can use analytics and get important value from it. In the 1950s, decades before anyone voiced the term "big data," businesses were using basic analytics (basically numbers in a spreadsheet that were manually observed) to discover insights and trends.

The assistances that big data analytics offer to the table are speed and efficiency. A few years ago, a business would have collected information, applied analytics, and extracted information that could be used for upcoming decisions, whereas today that business can find understandings for immediate choices. The ability to work quicker—and stay agile—gives organizations a competitive advantage they did not have before.

4.1.7 IMPORTANCE OF BIG DATA ANALYTICS

Big data analytics helps organizations to harness their data and use it to identify new opportunities. That, in turn, leads to smarter business moves, more efficient operations, higher profits, and happier customers.

Following are the reasons why big data analytics are crucial

- Cost reduction: Big data technologies such as Hadoop and cloud-based analytics carry important cost benefits when it comes to storing great amounts of data—and they can recognize more well-organized ways of doing business.
- Faster, better decision-making: With the speed of Hadoop and in-memory analytics, joint with the capability to examine new sources of data, businesses are able to study information instantly—and make choices based on what they have learned.
- New products and services: With the capability to measure customer needs and fulfillment over analytics comes the power to deliver what the customers want and need. Therefore, with big data analytics, more companies are making new products to satisfy customers' needs.

4.1.8 GENERATION Y (GEN Y)

Gen Y is the generation of people born during the 1980s and early 1990s. The term is built on Gen X, the generation that came before them. Members of Gen Y are often referred to as "echo boomers" because they are the children of parents born during the baby boom (the "baby boomers"). Because people born during this time period have had continuous contact with the technology (computers and cell phones) in their youth, they required many employers to up-to-date their hiring plan in order to include modernized forms of technology. They are also called millennials, echo boomers, Internet generation, iGen, and net generation.

4.1.9 GENERATION Z (GEN Z)

Also known as the Net Generation or digital natives, Gen Z were born approximately between 1990 and 2000. A generational theme to partly describe Gen Z, it is instantly and extensively interconnected and connected locally and globally. The founder of a Gen Y social networking site, Penelope Trunk, stated that Gen Z will open doors that are barely being knocked by the previous generations due to Gen Z's ability to process the amount of information that they have been able to process from an early age. Gen Z is emerging into a multigenerational workforce and civic community and the motto of Gen Z is work to live instead of live to work. Since Gen Z is in comfortable position in participating in an interconnected world, in using various technology and global-minded skills to solve problem. For Gen Z, the future is very bright and holds promise not only for traditional methods of engagement but for new technology beyond our imagination.

4.1.10 GENS Y AND Z ATTRIBUTES

Technology is perceived to be so significant within the loves of Gen Y that researchers have labeled new generations on the basis of their familiarity with digital devices. Prensky (2012) coined the term digital native to describe a generation of tech-savvy individuals with a positive attitude to technology, accustomed to the speed of the Internet and to multitasking. The "net generation" is goal and achievement oriented, with a preference for active learning and social activities.

One study defines Gen Y as intelligent, ambitious, questioning individuals accustomed to networking, multitasking, always connected, and often more academically qualified than their managers, who are expected to be mentors and coaches.

Some have suggested that Gen Y seeks personal fulfillment, a meaningful job in a friendly organization, work–life balance and is also strongly motivated by career advancement (Meierand Crocker, 2010). Finally, according to research by PwC (Finn and Donovan, 2013), unlike Gen X, they prioritize support, appreciation, and flexibility over salary.

4.1.11 GENS Y AND Z EMPLOYMENTS

According to the latest ONS figures (2014), over 3 million young people (aged 16–24) are part of the UK workforce, compared with 733,000 unemployed. There are also 2.72 million young people economically inactive, mostly full-time students who will join the workforce soon.

Now, the three-quarters of employers offer programs aimed at young people, the most popular of which are work experience and apprenticeships. Graduate schemes, apprenticeships, and school-leaver programs are considered the most effective pathways to work.

KEY PROGRAMS

The key programs are as follows:

- Industrial placements
- Work-based degrees
- Apprenticeships
- Graduate programs

4.2 LITERATURE REVIEW

The scientific research has been revolutionized by "big data." As technology advances, particularly with the advent of next-generation sequencing, the size and number of experimental datasets available are increasing exponentially.

"Big data" has the potential to revolutionize not only research but also education. A recent detailed quantitative comparison of different approaches taken by 35 charter schools in New York City has found that one of the top five policies correlated with measurable academic effectiveness was the use of data to guide instruction. Imagine a world in which we have access to a huge database where we collect every detailed measure of every student's academic performance. This data could be used to design the most effective approaches to education, starting from reading, writing, and math, to advanced, college-level courses. We are far from having access to such data, but there are powerful trends in this direction. In particular, there is a strong trend for massive web deployment of educational activities, and this will generate a large amount of detailed data about students' performance.

It is widely believed that the use of information technology can reduce the cost of healthcare while improving its quality by making care more preventive and personalized by extensive (home-based) continuous monitoring.

4.3 RESULT AND DISCUSSIONS

4.3.1 UNITED PARCEL SERVICE (UPS)

United Parcel Service (UPS) is the world's main package delivery company and a leading global provider of specialized transportation and logistics services was founded in 1907. This company provides a commanding competitive advantage that can help earn repeat customers and grow their business by using advanced technology, access to global resources, and an integrated network of physical, technological, and human assets.

In 1907, there was a great need for private messenger and delivery services in the United States. An innovative 19-year-old, named James E. Casey (also known as Jim), borrowed 100 dollars from an acquaintance and started the American Messenger Company in Seattle, Washington to help satisfy this need. According to Jim, there were quite a number of messenger services already in Seattle, some of which he had already worked for in the past years. In response to receiving telephone calls at their basement headquarters, messengers ran errands to deliver packages and carry notes, baggage, and trays of food from restaurants. Most deliveries were made on foot, whereas bicycles were used for long distance trips. There were only a few automobiles available at that period of time and the department stores still used horses and wagons for goods delivery. After 6 years, only then the United States Parcel Post system was established. And today, UPS manages more than 16 million shipments a day, all over the world.

4.3.2 UPS IN GEN Y (1980–2000)

Year	Happenings
1975	UPS went universal for the first time when it offered services and administrations within the Canadian city, Toronto.
1980s	The demand for air parcel delivery expanded and federal deregulation of the aircraft business created new openings for UPS.
1985	UPS started international air service between the United States and six European countries.
1989	Then domestic air service was added in Germany.
1993	UPS was delivering 11.5 million packages and documents a day for more than 1 million regular customers. With such a huge volume, UPS had to develop new technology to maintain efficiency, keep prices competitive, and provide new customer services. Technology at UPS spans an incredible range, from small handheld devices to specially designed package delivery vehicles and global computer and communications systems. The handheld Delivery Information Acquisition Device (DIAD), carried by every UPS driver, was developed to immediately record and upload delivery information to the UPS network. The DIAD information even includes digital pictures of a recipient's signature, thus giving customers real-time information about their shipments. Furthermore, this proprietary device allows drivers to stay in constant contact with their headquarters, keeping abreast of changing collection schedules and traffic patterns and receiving other important messages. At the other end of the spectrum, UPSnet is a global electronic data communications network that provides an information processing pipeline for international package processing and delivery. UPSnet uses more than 500,000 miles of communications lines and a dedicated satellite to link more than 1300 UPS distribution sites in 46 countries. The system tracks 821,000 packages daily.
1986–1991	UPS spent USD 1.5 billion on technology improvements and plans to spend an additional USD 3.2 billion over the next 5 years. These investments are aimed at improving efficiency and expanding customer service.
1992	UPS began tracking all ground packages.
1994	UPS.com went live, and consumer demand for information about packages in transit soared. The following year, UPS added functionality to its website so customers could track packages in transport. The resulting popularity of online package tracking exceeded all expectations.

Year	Happenings
Late 1990s	UPS was in the midst of another transition. Although the core of the business remained the distribution of goods and the information that accompanies them, UPS had begun to branch out and focus on a new channel: services. As UPS management observed, the company's expertise in shipping and tracking positioned it to become an enabler of global commerce and a facilitator of the three flows that makeup commerce: goods, information, and capital. To fulfill this vision of new service offerings, UPS began strategically acquiring existing companies and creating new kinds of companies that did not previously exist.
1995	Through these acquisitions and creations, UPS sought to serve its customers in a new way. By providing unique supply chain solutions, UPS allowed its customers to better serve their own customers and focus on their core competencies. UPS formed UPS Logistics Group to provide global supply chain management solutions and consulting services based on customers′ individual needs. UPS acquired a company called SonicAir, making UPS the first company to offer same day, "next flight-out" service and guaranteed 8 a.m. overnight delivery
1998	UPS capital was founded with a mission to provide a comprehensive menu of integrated financial products and services that enable companies to grow their business.

4.3.3 UPS IN GEN Z (2000–2020)

Year	
2001	In 2001, UPS ventured toward retail business by acquiring Mail Boxes Etc., Inc., the world's largest franchisor of retail shipping, postal, and business service centers. Within 2 years, approximately 3000 Mail Boxes Etc. locations in the United States re-branded as The UPS Store® and began offering lower UPS-direct shipping rates. The stores remain locally owned and operated, and continue to offer the same variety of postal and business services, with the same convenience and world-class service
2003–2005	UPS continues to expand service worldwide. In Europe, Asia, and South America, customers enjoy an unmatched portfolio of time-definite and supply chain services. Two major enhancements to international service came with the expansion of Worldport, the air hub in Louisville, Kentucky, as well as the European air hub in Cologne, Germany. With Asia identified as a primary growth target, in 2005 UPS launched the first nonstop delivery service between the United States and Guangzhou, China. That same year, UPS acquired the interest held by its joint venture business partner in China, giving it access to 23 cities that cover more than 80% of the country's international trade.

Year	
	From using electric vehicles in New York City during the 1930s to developing water conservation techniques while keeping the familiar brown package cars clean, as well as operating the world's largest fleet of compressed natural gas vehicles, UPS has long practiced environmentally conscious innovations. Although sustainable practices are not new to UPS, the company recognized the need to formally document its focus on responsible business models. In 2003, UPS issued its first corporate sustainability report, highlighting the importance of balancing economic, social, and environmental objectives. Now due to an annual report, it tracks the company's key performance indicators relevant to the business.
	UPS continually gains wider access to various markets through acquisitions. The 1999 acquisition of Challenge Air made UPS the largest express and air cargo carrier in Latin America. Purchasing Menlo Worldwide Forwarding in 2004 added heavy air freight shipment capability, while the acquisition of Overnite in 2005 expanded the company's ground freight services in North America. Other recent acquisitions in the United Kingdom and Poland present new opportunities for growth in Europe.
	Over the past 100 years, UPS has become an expert in transformation, growing from a small messenger company to a leading provider of air, ocean, ground, and electronic services. The most recent public change came in 2003, when the company introduced a new brand mark, representing a new, evolved UPS, and showing the world that its capabilities extend beyond small package delivery. The company went another step further, adopting the acronym UPS as its formal name, another indicator of its broad expanse of services. Ever true to its humble origins, the company maintains its reputation for integrity, reliability, employee ownership, and customer service. For UPS, the future promises even more accomplishments as the next chapter in the company's history is written.
	Since becoming a publicly traded company in 1999, UPS has significantly expanded the scope of its capabilities primarily through the acquisition of more than 40 companies, including industry leaders in trucking and air freight, retail shipping and business services, customs brokerage, finance and international trade services. As a result, UPS's relationships with many of its customers have deepened to include much more than basic transportation services.
	Today, UPS operates an international small package and document network in more than 185 countries and territories, spanning both the Atlantic and Pacific oceans. With its international service, UPS can reach over four billion people, twice the number of people who can be reached by any telephone network. Moreover, UPS.com receives millions of online tracking requests daily.

4.3.4 BIG DATA IN UPS

UPS has a single integrated and optimized network which results in environmental benefits such as reductions in fuel use and greenhouse gas emissions. To make the network operate more efficiently and reduce environmental impact, UPS designs, acquires, implements, and optimizes information technology for continuous improvement. Technology has helped UPS fine-tune various aspects of its operations—from planning and routing to flying and driving—something that is good for business and the environment. A key tool in achieving sustainability improvements is the use of "big data." UPS uses proprietary package movement technology to decide what packages are filled on each transportation, then gathers data from numerous aspects of fleet processes using a telematics technology system.

The business records the performance and condition of their transportation engines. The drivers' behavior and safety habits on the road are recorded by implying GPS. Sensors are used to produce a report on the emissions and fuel consumptions. They also monitor deliveries and customer services through devices. Last, Maps are used to collect address points and routes covered.

These technologies and sensors capture over 200 data points from more than 80,000 transportations every day. The data analyzed includes

1. Engine monitoring
2. Speed
3. Mileage
4. Number of stops
5. Miles per gallon
6. Safety aspects

4.3.5 BENEFITS OF BIG DATA IN UPS

By using these data and technologies, UPS has gained exponential outcomes. These developments have aided the business find ways to sort and load packages more precisely and accurately. These advancements also allow a company to offer drivers a customized manifest of packages on each vehicle to guarantee optimized delivery. These efforts drive efficiency and service improvements, as well as environmental benefits.

4.3.5.1 FUEL AND EMISSIONS EFFICIENCY

In 2012, the amount of idling time of 206 million minutes was reduced by UPS, which led to the saving of more than 1.5 million gallons of fuel.

4.3.5.2 GREENHOUSE GAS REDUCTION

UPS also avoided 13,000 metric tonnes of carbon emissions due to reduced miles per stop. This has resulted in the removal of 12.1 million miles of driving in 2012.

4.3.5.3 FUEL REDUCTION

About 39 million gallons of fuel were saved since 2001 through route optimization and monitoring.

4.3.5.4 MILEAGE REDUCTION

Lastly, 364 million miles of driving time has been avoided by UPS since 2001.

4.3.6 ON-ROAD INTEGRATED OPTIMIZATION AND NAVIGATION (ORION)

As UPS has numerous parts and pieces moving around the world, they store a large amount of data which majority are coming from sensors in their transporting vehicles. These collected data not only analyzed daily performance and activities but also triggered a major restructure of UPS drivers' route structures. The ingenuity was called On-Road Integration Optimization and Navigation (ORION). This system depended heavily on data gained from online map to readjust a driver's pickups and drop-offs in real time.

UPS takes an all-inclusive method to integrate data into technology. ORION was arguably the world's largest operation research project. ORION uses fleet telematics and advanced algorithms to take route optimization to a new level. In 2013, UPS started the first major deployment of ORION, with plans to deploy the technology to all 55,000 North American routes by 2017.

4.3.6.1 OUTCOMES OF ORION

1. Saved 1.5 million gallons of fuel savings and 14,000 fewer carbon dioxide emissions by end of 2013 by optimizing 10,000 routes
2. Collected more than 250 million address data points
3. Planned 55,000 North American routes for deployment by 2017
4. Has 10,000 of route optimizations per-minute based on real-time information

4.3.7 LEARNING AND DEVELOPMENT FOR GENS Y AND Z

When young people enter an organization, it is important to provide a training and developing programs personalized to their exact needs, instead of a common one (Oxenbridge and Evesson, 2012, Culliney and Broughton, 2013). The induction is an opportunity to meet colleagues, to familiarize them with the new situation, together with being introduced to both the company and their specific role (Culluney and Broughton, 2013). While this is valid for all employees, it is especially relevant for youngsters, as the shift from education to employment might be "traumatic and disorienting" and many lack confidence (Smith 2003, p17, Oxenbridge and Evesson, 2012).

Young people also specifically benefit from clear objectives, regular feedback, opportunities for upward communication and guidance on the skills they can develop or are developing on the job.

Further, they need a strong IS support network, according to most employers. Suggestions include mentoring and coaching support from an effective line manager (Peate and Taylor, 2014). Buddies or other peer support can also be particularly helpful for young people, who may feel less at ease with their manager initially.

A range of learning and development methods is used to support ongoing development. For example, individual- and group teaching-guided study and social learning are frequently adopted in apprenticeship programs (Peate and Taylor, 2014). Collaborations are often formed to offer thisgeneration, especially Gen Z, the chance to work and earn formal qualifications. For example, there has been an increase in joint partnerships between private companies and universities to deliver work-based degrees (UK Commission for Employment and Skills and Universities UK, 2014).

4.3.8 LEARNING METHODS FOR GENS Y AND Z

4.3.8.1 LEARNING FROM EXPERIENCE

- On-the-job learning: one of the primary development methods used is learning from experience, on the job. For example, ActionAid is focused on enabling apprentices to learn from experience.
- Business problem exercises: used by a number of organizations to help people gain experience of working on a project, data analysis, and completing different challenges.
- Giving presentation: other generations have also used presentations as a development method, particularly, as presenting to senior leaders is a key way of gaining real-life experience of a challenging situation.

4.3.8.2 LEARNING FROM OTHERS

- Mentoring: Mentoring using different forms is frequently used to build capability. A reverse mentoring also means that trainees mentor more senior colleagues on the business models they have learned as part of their higher education-based learning.
- Networking: It is commonly viewed as a helpful development technique, networking helps young generations build a network across the organization.
- Learning from peers and your line manager: While these techniques are actively managed, informally learning from your line manager and from the team is still a vital part of growing and development.
- Formal learning

    a. Intensive inductions: A quality induction period is very important as entering employment for the first time can be difficult for young generations. This period can help young people feel at ease and better understand their role and the organization. It is designed to help them make the transition back to work and help them build interpersonal skills.

"It helps them understand what it means to communicate effectively, to understand team working, how to plan and a whole range of core employment skills." Mike Thompson, Barclays.

b. Formal qualifications: Formal qualifications are used to enhance and build the individual's technical knowledge base, either as part of a work-based degree or through an apprenticeship. In an industry that places great value on having a degree, this also give apprentices the chances to be able to continue to progress and develop their career.
c. Classroom training: Young generation on the various schemes typically have access to a range of classroom training courses, which are available throughout the organization.

4.3.9 GENS Y AND Z LEARNING PREFERENCES

4.3.9.1 MIXED VIEWS ON LEARNING TECHNOLOGY

As highlighted in the introduction, many young people are "tech-savvy" and therefore typically use technology in their day-to-day lives, and make use of online tools to develop their knowledge. Many young people which were interviewed agreed with this perspective, and explained that they regularly use the Internet to access information—a form of self-directed learning. Especially for Gen Y who has confidence, they have high expectations, particularly when it comes to learning. In terms of learning, Gen Y will give more honest responses and due to this confidence that they are able to contribute to the class or lecture.

4.3.9.2 TECHNOLOGY IS ALSO USED AS A NETWORKING TOOL

"The use of technology between generations is probably the starkest difference… We have got a very active thriving online community, but when you look at the profile of people using it, it does tend to skew more toward the younger generation." Mike Thompson, Barclays.

These findings reflect the preferences young people exhibit for using technology, and the "tech-savvy" skills outlined in the previous part. However, there are a lot of significant challenges when this information is applied to learning methods. In Fujitsu, the young people shared their frustration with unwieldy learning technology and compliance e-learning. While young people may be tech-savvy, it represents a genuine challenge for the learning and development profession as we continue to embed the use of technology for learning.

4.3.9.3 PREFERENCE FOR LEARNING FROM DOING

Many of younger generations interviewed expressed a desire to learn from experience and working on collaboration with others.

Niger Spencer, Reed Smith, reflects on the value of on-the-job experiences on the wider transition young people face

> What really makes the difference in going from the world of study to the world of work is the outside-the-classroom experiences. That's why I was so passionate about putting in place things like the client intelligence projects and the reverse mentoring as they areworkplace-based, real situations in which to apply their learning.

4.3.9.4 BITE-SIZED LEARNING

Traditional classroom training may not always suit young generation's preferences especially if they have recently left school.

4.3.9.5 MAKING A DIFFERENCE AND GETTING FEEDBACK

Gaining feedback, not just from senior managers but also from peers was viewed as a great way to understand the difference that you are making and how you can learn and develop.

4.3.10 INDUSTRY 4.0—EVERYBODY AND EVERYTHING IS NETWORKED

4.3.10.1 CHANGING DEMANDS: QUALIFICATION AND TRAINING FOR INDUSTRY 4.0

Industry 4.0 is also known as the fourth industrial revolution, which will address and solve some of the challenges facing the world today such as resource and energy efficiency, urban production, and demographic change.

4.3.10.2 CHANGING EDUCATIONAL DEMANDS: NEW WORKING WORLD WITH NEW PLAYERS

- Technical world
  1. New phase of automation technology
  2. Robots outside fences/cages
  3. Mobile robotics
  4. Robots in everyday life
- Globalized organizations
  1. Structures above geographical distances
  2. Globalized virtual communication
- Organizational functioning
  1. New types of organizations
  2. New ways of performance and success measurement

4.3.10.3 NETWORKED WORLD LEADS TO NEW MODELS IN SCIENCE AND EDUCATION

However, in second revolution industry the resource of power is emphasized, in other hand the resource of intelligence is highlighted in fourth industrial revolution.

Some of the 21st Century Learning Topics

- Lifelong learning
- 24/7 access to information
- Multiple, hybrid

- Interdisciplinary nature
- Networked, independent
- Expansion of the skills
- Personalized
- Adaptive and flexibly
- Operations research
- Artificial intelligence
- Neurosciences
- Soft skills
- Dealing with diversity
- Language skills
- Communication
- Cooperation
- Changed roles
- Creativity
- Innovation
- Abstraction

4.3.10.4 CHANGING EDUCATIONAL DEMANDS LEADS TO NEW EDUCATION MODELS

Some of the 21st Century skillset

- Solving complex, multidisciplinary, open-ended problems within changing and partly anonymous teams
- Thinking critically and making judgments
- Creativity and entrepreneurial thinking
- Communicating and collaborating in new ways
- Making innovative use of knowledge and information

4.3.10.5 EMPLOYEES OF A NEW WORKING WORLD, INDUSTRY 4.0, MATCHING SUPPLY AND DEMAND FOR GENS Y AND Z

- Learning strategies/behavior of Gen Y and Z
- Gen Y focus on social interaction and collaboration, self-paced, tactile learning, that is, doing is more important than knowing,

whereas Gen Z focus on "Learning by playing," collaborative and connected learning and Internet as extension of one's self

- Changes in vocational education
  1. Subject-oriented systematic; use of established subject classification
  2. Process-oriented perspective; Integrating generic, social, and methodical skills
  3. Process orientation with reflection based on subject classification; focus
  4. Learning: work process perspective and focus on reflection: perspective on technology

4.3.11 EMPLOYEES OF INDUSTRY 4.0

4.3.11.1 MULTI- AND INTERDISCIPLINARY CHALLENGE

Orientation Toward Digitalization for Mechanical Engineering (ME)

- **Web-based application/solution**

Engineers, especially those who are doing mechanical works, can use web-based application to run a website or any web application that can help designing infrastructure or architectural design with the ease of access to IS.

Google Docs is an example of a decent web-based application that allow workers in this field to share their data collected in any web browsers, for example, Safari and Google Chrome to their fellow colleagues. By using this method, individuals can save much time and costs.

- **Decentralization**

In general, the decentralization process can substantially increase efficiency, transparency, accountability, and responsiveness of service provision compared with centralized system.

A decentralized system is characterized by the exercise of substantial power at the local level on many aspects of features education, subject to some limited control by the central government (World Bank, 1997).

- **Digital Trial and Error**

A method of reaching a correct solution of satisfactory result by trying out various means or theories digitally until error is sufficiently reduced or eliminated. For example, a trial-and-error approach to building an automated bridge will take a longer time to correct than building just a normal bridge.

- **No language/time barriers**

Using IS, virtually, you can setup meetings with fellow engineers, clients, managers, and so forth which can save ample amount of time of the workers. By using IS application, software or hardware, workers can no longer have barriers, or reduced to some extent, to language as the IS itself can help the engineers to understand better.

- **Internet of things (IoT)**

IoT link smart objects to the Internet. Never available before data can be exchanged and bring users information in a more secure way (Cisco, 2016). Cisco also predicts that IoT will consist of 50 billion devices connected to the Internet by 2020.

- **Cyber-physical systems**

A cyber-physical system (CPS) is a mechanism controlled by a computer-based algorithm, strongly integrated with Internet and its users. Examples of CPS are as follows: smart grid, medical monitoring, process control systems, robotics systems, autonomous automobile systems, and automatic pilot avionics.

- **System security**

Control of access to a computer system's resources, especially its data and operating system file

- **Smart data/big data**

"Smart data" means information that actually makes sense and its main purpose (veracity and value) is to filter out the noise and hold data which are valuable. These smart big data can be used effectively by engineers or businesses to solve complex problems.

4.3.11.2 SCIENTIFIC PROGRAMMING ALSO BECOMES A MAJOR PART OF ME

Virtual Production Environment (VPE): Virtual Production Environment is a model of a factory that is used by the a priori application to produce cost assessments (apriori, 2016) Information contained in a VPE:

a) Supported manufacturing process groups, processes, and routings
b) Manufacturing feasibility rules
c) Materials
d) Machines, machine capabilities, and constraints
e) Machine time standard information
f) Machine and tooling accounting rates
g) Process model logic
h) Cost model logic

Product Design: Product Design is a broad approach to the designing and making new innovative products which are to be sold by a business to its customers.

E-engineering: E-engineering offers new ways of managing complex problems that usually arise in the engineering world, but businesses should also be cautious of what unique issues this virtual work process brings. E-engineering can drastically cut technology costs for companies and by e-engineering, companies can gain competitive advantage due to the low costs involved using this process.

4.4 CONCLUSION

Having a determination of the right approaches, an awareness of generational learning preferences is essential. It is also clear that these generations, Y and Z, have high expectations as they are the regular user of these technologies and proficient enough to use it as a communication tool and to learn in the young people's personal lives.

This generation is divided between Y and Z, incorporate with the need to develop core competencies in compulsory education and also in businesses (especially digital/IS competence), adapt to new social skills related to the use of technologies and to explore the skills young generations should have and those they need to build. By taking different learning approaches, they can often identify areas for improvement and have the initiative to find solutions themselves. The needs of a changing society where everyone within the organization will bring every skills and strength acquired in order to have the chance to impact the organization at an early stage, raise questions about the preparation of current teachers for leading the teaching–learning processes that the future generation of students will use.

We proposed a deep learning approach about revealing big data for both Gen Y and Gen Z for learning process. For example, the 4th industrial revolution is already running and since it is an IT-based, connectivity is one of the main characteristics. The new working worlds demand new skills as new technologies change our views of working and behavior. The technological breakthroughs - like algorithms and computational power to work on these big data made promising qualification concepts to be available as learning analytics in augmented training scenarios which are projected to train and gain the needed skillset of the 21st century. However, it is best not to assume that IS or IT will always be the proper learning solution for Gens Y and Z.

KEYWORDS

- **Generation Z**
- **healthcare**
- **internet**
- **research**
- **united parcel service**

REFERENCES

Culliney, M.; Broughton, A. Young People's Views and Experiences on Entering the Workplace [online]. Research Paper 09/13. Acas: London. Available at: http://www.acas.org.uk/media/pdf/c/5/Young-peoples-viewsand-experiences-on-entering-theworkplace.pdf2013.

Finn, D.; Donovan, A. PwC's NextGen: A global generational study: evolving talent strategy to match the new workforce reality [online]. [London]: PwC. Summary available at: http://www.pwc.com/us/en/people-management/publications/nextgen-global-generational-study. jhtml [accessed 12 March 2015], 2013.

Meier, J.; Crocker, M. Generation Y in the workforce: managerial challenges [online]. *J. Hum. Resour. Adult Learn.* **2010,** *6*, 68–78. Available at: http:// www.hraljournal.com/ Page/8%20 Justin%20Meier.pdf [Accessed 18 March 2015].

Oxenbridge, S.; Evesson, J. Young People Entering Work: A Review of the Research [online]. Research Paper 18/12. Acas, London. Available at: http://www. acas.org.uk/ media/pdf/5/2/Youngpeople-entering-work-a-review-ofthe-research-accessible-version. pdf2012.

Peate, A.. Taylor, D. Managing Future Talent [online].CIPD, London. Available at: http:// www.cipd.co.uk/publicpolicy/policyreports/managing-future-talent. aspx2014.

Prensky, M. From Digital Natives to Digital Wisdom: Hopeful Essays for 21st Century Learning. Corwin: Thousand Oaks, CA, 2012.

UK Commission for Employment and Skills (UKCES) and Universities UK (UUK). Forging Futures: Building Higher Levels Skills Through University and Employer Collaboration. https://www.gov.uk/government/uploads/system/uploads/attachment_data/file/356749/ FF_FinalReport_ Digital_190914.pdf**2014**.

World Bankhttp://www1.worldbank.org/publicsector/decentralization/Feb2004Course/Background%20materials/Florestal.pdf1997

CHAPTER 5

CREATING TECHNOPRENEURS AND INNOVATION: A BIG DATA APPROACH

HERU SUSANTO[1,2,*] and CHIN KANG CHEN[3]

[1]*Indonesian Institute of Sciences, Jakarta, Indonesia, E-mail: heru.susanto@lipi.go.id*

[2]*Tunghai University, Taichung, Taiwan*

[3]*University of Brunei, Bandar Seri Begawan, Brunei Darussalam, E-mail: chinkang.chen@ubd.edu.bn*

[*]*Corresponding author. E-mail: chinkang.chen@ubd.edu.bn*

ABSTRACT

The main purpose of this chapter is to gain more knowledge and completely understand, analyze on how useful the big data is, and how technology becomes a necessity for an individual, especially for an organization to learn in this modern/technology era. The Internet era had created a huge amount of data available to an individual or any organization in the world. These data are considered as big data, which comprises databases that is huge to be managed by outdated database systems. Structured and unstructured data are included in big data. Those data are formatted and unformatted, respectively. Structured data are used in database management, whereas unstructured data are used for all types, such as social media and multimedia. Sensors and actuators are the other types of data that are embedded in an object.

Technology has developed and shaped the human life and it plays an important role in almost all aspects of modern life. In many countries, technology has become one of the important drivers of the economy.

This chapter focuses on big data that can create a technopreneur and the nature of research is a qualitative design of a secondary research and the data were collected through observation notes, Google reflection essays, journals, and books. This chapter begins with a quick introduction on technopreneurs, a data management system, and big data, which is followed by literature review, some important open issues, result and discussion of big data. Conclusion and recommendation are also presented in the last section of the chapter.

5.1 INTRODUCTION

Technopreneur is a combination of the terms "technology" and "entrepreneur." Technology means the application of knowledge, which deals with the creation for practical ends; whereas, entrepreneur is a person who manages and organizes any business with considerable initiative and risk. This term has been used since 1987 for a technology as an integral and key element in the transformation of good and services. Technopreneurs are entrepreneurs who are really into a business that involves high technology and make use of it to produce new and innovative products through a commercialization process where both technical and business skills are applied for potential technopreneurs.

The technopreneurs are the entrepreneurs who are technology savvy, creative, innovative, dynamic, dare to be different and take the unexplored path, and very passionate about their work. They take challenges and strive to lead their life with greater success. They do not fear to fail. They take failure as a learning experience, a stimulator to look things differently and stride for next challenge. Technopreneurs continuously go through an organic process of continual improvement and always try to redefine the dynamic digital economy (Susanto, 2016a).

A *database management system* makes it possible for an end user to create, read, update, and delete data in a database. The important role of database management system is to serve as an edge between the folder and final client or purpose programs and ensure that the data is constantly planned and remains accessible.

A database administration system manages three significant things: the data; the database engine that allows data to be accessed, protected, and customized; and lastly the database schema, which defines the database's rational arrangement. These three initial fundamentals facilitate to provide concurrency, safety, data reliability, and standardized management measures. Classic database management responsibilities are supported by the database organization method that includes altering organization, routine monitoring or tuning, as well as support and improvement. Database organization system is accountable for computerized rollbacks, restart, and improvement, as well as for sorting and audit action.

With the advancement of technology in the past few decades, several popular database management structures are included. Relational database organization structure is flexible in most use cases, however, RDMS Tier-1 products are fairly costly. Second, the NoSQL DBMS, which is fit matched for a freely distinct data structure can develop the future point in time. Another database organization structure that is popular is the in-memory database management system, which provides quicker reaction period and enhanced act. Fourth, the columnar database organization arrangement is compatible fora data warehouse that has a huge quantity of similar data items. Lastly, in a cloud-based data organization structure, the cloud service provider is accountable to provide and maintain the database organization structure.

The above mentioned data and database management system have evolved with the explosion and breakthrough of modern technology and sciences in the arena of data management. Companies and business nowadays have the ability to store, process, and manipulate data in a manner that having a large pool of data of their clients, customers, and other stakeholders, which are readily available online and off-line, both in a digitized or physical state. These sets of data are no longer saved or stored into paper or cloth like the early human civilization.

As time progressed and more advancement in technology, data and information have been stored into laser disk, floppy disks, magnetic tapes, Universal Serial Busdrives, compact disks, hard drives, and others. Now, it is time for big data to fully use its potential in terms of data management capability to help companies and organization to gather information and process data that have not been used previously due to limitations of the older technology. However, with the ability to store huge amount of data and infrastructure such as cloud computing, we are able to stock up a limitless amount of information. Data sizes from terabytes and petabytes are being generated, processed, stored, and managed without difficulty.

"Big data" concept was initially introduced to the computing world by Roger Magoulas from O'Reilly Media in 2005 to describe a huge quantity of data which conventional data organization technique cannot deal with and process due to the difficulty and amount of this data (Ularu et al., 2012). They also continued to state that, according to MIKE 2.0, the open resource typical for information management, big data is distinguished by its amount, comprises a big, multipart, and autonomous collection of data sets, each with the possibility to interact. In this, each big data is capable of making communication two-way or multiple-ways.

Big data can be distinguished as a set of data sets that is huge and multipart which has become hard to process by means of conventional database organization paraphernalia and data processing application. In addition, it can consist of a huge volume of mixed data that is being generated at fast speed. These data sets cannot be managed and processed using conventional data organization paraphernalia and application. Big data requires the utilization of the latest set of tools, application, and frameworks to process and handle the data.

Based on the executive report processed by IBM Institute for Business Value in collaboration with Saïd Business School, University of Oxford (2012), big data is preeminently described as today's superior quantity of data, the latest type of data and study, or the up-and-coming necessities for more concurrent information study. It is further stated that big data are characterized by several features. It is a better extent of information, it is a new kind of information and analysis, it shows immediate information, data gathered are in flood from new technologies, use nontraditional forms of medium, uses huge volume of data, it is the newest catchphrase, and uses community medium data.

Another meaning of big data affirmed by Susanto (2018) refers to information sets or combinations of information sets whose amount (volume), difficulty (variability), and pace of expansion (velocity) make them hard to be captured, managed, processed, or analyzed by conservative technologies and tools such as relational database and desktop figures or visualization parcels, within the time needed to make them functional. The amount used to decide whether a meticulous data set is considered big data is not definitely defined and continues to alter over time. At present, most analysts and practitioners refer to information sets from 30–50 TB (1012 or 1000 GB/TB) to several petabytes (1015 or 1000 TB/PB) as big data.

Most people consider big data as a new revolution in database management system but in reality the main core for big data—the information itself has been around. On the other hand there has always been a need for storage space, dispensation, and the organization of data since the beginning of human evolution and human society. However, during primeval times, data and information that have been gathered would be chosen depending on its significance. The quantity and type of data captured, stored, processed, and managed depend on various factor such as requirement, present technologies, ease of using tools for storage, processing, organization effort or cost, the ability to gain insight into the information, the ability to make judgment based on the data, and other functions.

As stated by Susanto (2018), a study on *the Evolution of Big Data as a Research and Scientific Topic* shows that the term "big data" was present in the study from the 1970s but was introduced in publication in 2008. They also stated that nowadays the big data concept is treated as diverse point of view, covering its implication in many fields.

Big data has completely transformed the way businesses and organizations operate. It has an impact on majority of the world population without realizing it. Big data is also defined as ever more send-off a digital outline or information, which can be used by us and others and analyzed. It is our capacity to create and make use of the ever-growing volume of data.

As stated by Eric Schmidt, Executive Chairman of Google, "from the dawn of civilization until 2003, humankind generated five exabytes of data. Currently, we create five exabytes every two days and the pace is accelerating." This shows that the huge quantity of information that we are producing everyday needs a structure or ways to manipulate the information produced and change it into useful information. With big data, it is possible to carry out these tasks.

5.2 LITERATURE REVIEW

There are many previous study and research related to creating technopreneur and innovation through big data. Each study had its own view about technopreneur, innovation, and big data. As the Internet changed, as well as the information technology (IT) spread faster, most of the data were born digitally. Although the advancement of computer systems and Internet technology have renewed the computing hardware for years, as we entering the age of big data, handling and controlling large-scale data is still one of the problems that never ends.

The article by Bendre and Thool (2016) has stated the definition of the big data, where by using old-fashioned data processing system one cannot manage and support the large quantity of data. They introduced the background and anything that was related to the big data and technologies. Big data analysts also observed and stated that big data analytics is a process of investigative information and patterns from large data in their article and Shu (2016) also gave different opinions about big data and big data analytics. He has stated that big data is defined by four *Vs*—these are volume, variety, velocity, and veracity. The six techniques are also introduced for the big data analytics that related to the four *Vs*, which include collective analysis (related to the volume), association analysis (data sampling), high-dimensional analysis (variety), deep analysis that connected to veracity of big data, precision analysis (veracity), and lastly, divide-and-conquer analysis (veracity).

The use of big data is to better understand the behavior and preferences. As Dalton (2016) stated in his article about the prospective of big data of political behavior. The essay that he discussed is the limitations of big data and current potential of political behavior. It is still uncertain whether big data can well address the academic and theory testing research challenges. Jenkins et al. (2016) have introduced an article titled *Political Behavior and Big Data.* The usage of big data nowadays has become dramatically higher. It has various meanings in the applications of big data and may have a deep understanding of political behavior.

Pak (2012) signified the importance of the innovation and entrepreneurship in Singapore. He analyzed on how the universities and school can develop innovation and entrepreneurship. There are two important ingredients to the continuous economic.

Chong and Shi (2015) pointed that big data has become a big challenge to process if huge amount of data is generated. Big data is used to expose unknown information and achieve competitive advantage in the market but unfortunately, huge publication of big data analytics makes it problematic for scholar and any other researchers to search matters. To solve this problem, organization needs to find new methods and tools for the process of big data.

5.3 RESULT AND DISCUSSION

5.3.1 BIG DATA DATAFICATION

5.3.1.1 ACTIVITY DATA

Marr (2015) stated that reading books at e-books and listening to music by digital music players are now generating data. Smartphones collect data and web browser collects information on how much we use it or what we are searching on it. A credit card company collects data regarding what and how much we buy.

5.3.1.2 CONVERSATION DATA

Based on the study of Marr (2015), it refers to any conversation that we made, which leaves a digital trail, for example, conversation that we made on WhatsApp, Facebook, or Twitter and any other social media. Even in our phone, conversation data are digitally recorded.

5.3.1.3 PHOTO AND VIDEO IMAGE DATA

Based on the study of Marr (2015), photo and video image data refer to the pictures and videos we take by our smartphones and cameras and share them every second on social media sites. The video images uploaded on YouTube that had been taken by closed-circuit television (CCTV) are also increasing the amount of data.

5.3.1.4 SENSOR DATA

Based on the study of Marr (2015) and Susanto (2017c), it refers to data collected and shared by sensors. For example, every smartphone now provides sensor to track where we travel and to track the speed or direction accelerometer is included.

5.3.1.5 THE INTERNET OF THINGS DATA

Based on the study of Marr (2015), it refers to the data that have been collected and processed by smart televisions, watches, refrigerators, and alarms. The Internet of Things or Internet of Everything connects these devices so that traffic sensors on the road send data to our alarm clock which will make us wake up earlier than planned because blocked road will mean that we have to leave early to be on time. Thus, it is interrelated and functions really well as long as the data collected are accurate.

5.3.1.6 BIG DATA FIVE VS

Volume: as the amount of data that a company gathered. It is used to collect important information.

Marr (2015) defines it as the vast amount of data generated every second. The data are generated in zettabytes or brontobytes. For instance, before 2008, data have been produced in the same amount in the world and is soon generated every minute. Thus, by using traditional database technology, it makes data sets huge to analyze and store. We can store, retrieve, and analyze data across databases that are scattered anywhere around the world by using new big tools that used to distributed systems.

According to Gupta et al. (2014), volume refers to the large size of data than terabytes and petabytes, which make it difficult to store by using old style tools. For example, Facebook ingests 500 TB of data every day.

IBM institute for Business Value in collaboration with Saïd Business School, University of Oxford (2012) stated that one characteristic of big data that is most associated with its volume is to improve decision-making for an organization. The volume of data is continuously increasing at an unusual rate. It is smaller than petabytes and zettabytes and found that high volumes varied by industry.

Velocity: refers to the time to process big data. Some activities need response immediately and it is very important to process faster and efficiently.

Another definition by Marr (2015) is that velocity is considered as a speed of both new data generation and data movement. For instance, social media spread faster and going viral at any time, any minute, and any second. While velocity is being produced, technology nowadays allows us to examine the data without ever placing it into databases and sometimes, it referred to as in-memory analytics.

According to Gupta et al. (2014), velocity is needed for all processes. In order to maximize its value, big data should be used to the full stream of data. This is explained for time-limited process.

Another explanation by IBM Institute for Business Value in collaboration with Saïd Business School, University of Oxford (2012) is that the data that is created, examined, and processed is referred to the data in motion and speed which continues to accelerate. The real-time nature of data formation and the need to include streaming data into business procedure and decision-making are the factors that contribute to the higher velocity. It is further stated that the lag time between data is taken or produced or created when the data is available. It is where the velocity affects latency. Nowadays, traditional systems are unable to capture, examine, analyze, or store data where the data is continually being produced or generated at a pace time. Moreover, some categories of data must be evaluated in real time to be of value to the business. This is called time-sensitive processes, for example, multichannel "instant" marketing or scam detection.

Variety: variety of big data is a data that can be structured or unstructured.

Marr (2015) defines variety as diversity. It means that there are various types of data that we use nowadays such as text, images, voice, videos, and others. Nearly 80% of the world data is unstructured but now we can analyze data and bring different types of data together by using big data technology such as sensor data, social media conversation, and many more, which is different from the past that only focused on structured data and neatly fitted into relational databases and tables.

Gupta et al. (2014) stated that there are varieties of sources that big data come from. Big data is also geospatial data, audio, video, 3D data, and unstructured text, including social media; whereas, traditional database system was intended for a smaller volume of structured data, consistent data structure, and fewer updates.

Another explanation by IBM Institute for Business Value in collaboration with Saïd Business School, University of Oxford (2012) is that variety contains both data types and data sources, which handle the complication of numerous types of data such as structured, semi-structured, and unstructured. Nowadays, some organizations need to produce and analyze data from different array of both traditional and modern information sources within and outside the organization. Data is produced in different forms—there are text, sensor data, web data, audio, tweets, video, click streams, log files, and others—that is, the advancement in sensors, social collaboration technologies, and smart devices.

Veracity: that it is the degree that a leader believes the used information for decision-making. Therefore, it is very important for the business future to get the right relationships in big data.

Marr (2015) defines veracity as the disarray or trustworthiness of data. Quality and accuracy are less manageable with numerous forms of big data. Big data and analytics technology allows us to work with data such as hashtags, typos, abbreviations, colloquial speech, as well as the accuracy of content. The volumes frequently make up for the absence of quality or accuracy.

According to IBM Institute for Business Value in collaboration with Saïd Business School, University of Oxford (2012), veracity refers to data uncertainty, that is, the level of consistency related with specific types of data. It is very important for big data user to strive for high data quality as it is a condition. However, for the inherent unpredictability of some data cannot be removed by the best data cleansing tools such as weather. There is a need to acknowledge strategy for hesitation as one of the features of big data.

Value: The value means the ability to turn big data into value or meaning,that is, the ability to access big data by major companies has enabled them to produced amazing value from their big data. The latest software and analysis methods have allowed users to influence all types of data to increase insights and enhance value organized with the latest technology such as distributed systems and cloud computing (Marr, 2015).

Susanto (2018) pointed that *Vs* are not sufficient to describe big data that we face now. Van Rijmenan also explained important aspects of big data and a big data strategy that every organization cannot avoid. He also claimed that 13 years ago, META Group (known as Gartner) wrote a report about data management challenge. The title was "3D data management: Controlling Data volume, Velocity and Variety." In 2014, added more *Vs* to make some complement explanation about big data.

Validity: According validity of big data is defined as a quality of data, heterogeneous, and also as "unclean" data collection. Susanto (2018) defined validity as a correct and exact data for the planned purposes. Clearly, valid data is the solution to make the right choice.

Variability: has defined variability as the study of an object's non-static deportment in data sources, and also known to develop gradually.

Variability is also known as drastic change as well as change in performance of sentiment analyses. It is similar to using the same thing every day, but each day it feels and smells differently.

Venue: As we know, venue is location. For this big data, venue is the location that affects access to the data. Venue in big data is distributed. It is varied data from multiple stages, from different holder's system (owners) with different access and configuring requirements between private and public cloud.

Vocabulary: defined vocabulary as the data models that combine data from different sources with different names. He called it metadata, which describesthe data structure, content, and provenance.

Vagueness: Vagueness is a confusion about the meaning of big data. The big data at present is still unclear but it is also making the big data precise, clear, and convinced.

5.3.1.7 IMPORTANCE OF BIG DATA

There are specific areas or parts that big data can be used efficiently. First, examining the patterns to improve security and troubleshooting in information system technology. Second, customizing services in order to get customer pattern and improve customer satisfaction by using data from call centers in customer service. Third, by improving the services and products by knowing the possible customer's or consumers' preferences, the organization can change its product with the help of social media in order to address more people in area. Fourth, by analyzing and examining information from the transactions on the financial market in the risk assessment, and by detecting scam or fraud in online transactions for industry.

5.3.2 BIG DATA SOURCES

5.3.2.1 TRANSACTIONAL DATA

It is the data that is generated based on transaction. The transaction is an arrangement of information exchange which is related to the task such as database updating that is treated as a unit for the purposes of satisfying a request. Transactional data can be financial, logistical, or work-related, which involve everything that begins with purchase order, shipping status, hours an employee worked, insurance cost, and claims. It is grouped with both master and reference data. It also records a time and relevant reference data needed for a particular transaction record.

5.3.2.2 SOCIAL MEDIA

Social media such as Facebook and Twitter can generate numerous information and data from the comments and tweets. This data can be captured and analyzed to understand, for example, what people think about new product that is introduced.

5.3.2.3 ACTIVITY GENERATED

The record of any user action—online or in the physical world—that can be logged on a computer. It can be grouped into three categories. First, access, that is, logs of user access to systems indicating where users have travelled. For example; login and logout, passing through routers and other network devices or premises access turnstiles. Second, attention—it means that navigation of applications indicating where users have been paying attention, for example, page impressions, menu choices, and searches. Lastly, activity—"real activity," records of transactions which indicate strong interest and intent, for example, purchases, event bookings, lecture attendance, book loans, downloads ratings, and others.

5.3.2.4 PUBLIC DATA

Public data refers to the information that can be freely used, reused, and redistributed by anyone with no existing local, national, or international legal restrictions on access or usage. For example, in the enterprise, data can be classified as public if the information is available to all employees and all individuals or entities external to the corporation. The examples of public data in the enterprise include press releases, job descriptions, and marketing materials intended for the general public.

5.3.2.5 ARCHIVES

An archive is a collection of data transferred to a repository for backup to keep separate for compliance reasons or for moving off primary storage media. It can include a simple list of files or files organized under a directory or catalog structure, depending on how a particular program supports archiving.

5.3.2.6 FORMATS OF DATA

Big data is made of both structured and unstructured data information associated with the data type that may identifiable based on it is organized in a homogenous structure or heterogeneous of its type. The following subsection explained deeply on the data type and its structure.

5.3.2.7 STRUCTURED DATA

Syedet al. (2013) stated that this is the data where exact information is stored in columns and rows on a methodology that is also known as database. Structured data is also searchable by data type within content. Structured data is understood by computers and is also effectively prepared for human readers. The examples of structured data is spreadsheets and relational databases.

5.3.2.8 SEMI-STRUCTURED DATA

It is unstructured data that can be put into a structure with the help of available format description.

5.3.2.9 UNSTRUCTURED DATA

Syed et al. (2013) stated that unstructured data has no identifiable structure. It is defined as the information that does not have a predefined data model and/or does not fit well into relational tables. It is made up of 90% of Big Data and has "human information" such as e-mails, Facebook posts, call center conversations, Google, videos, CCTV footage, mobile phone calls, tweets, website clicks, and others. Within 2 years, big data got bigger up to 90% in the world. Software program is used to create the data and unstructured data, which contain its own specific structure because the data sources do have a structure but all data within a data set will not contain the same structure (Susanto, 2018; Almunawar, 2018b; Susanto et al., 2016; Almunawar et al., 2015a).

5.3.2.10 BIG DATA ADOPTION PROCESSES

Big data adoption processes can be classified into four main processes. It is important for every business and organization to carry out these processes to ensure the effectiveness and smooth running of the big data application and technology that can be used in the future.

Educate: Similar to the introduction of new technology or advancement, the first step to promote in the organization is to ensure all levels of management from top, middle, and lower management to be aware and have knowledge on the development of big data. This is important for an organization to use and apply big data tools or technology; the employees and stakeholders of the company need to know what the big data is all about. At this stage, the companies or organizations are focusing on knowledge gathering and market observation.

According to IBM Institute for Business Value in collaboration with Saïd Business School, University of Oxford (2012), people are not using big data within the organization as 25% of respondents remain ignorant about big data and technologies. Thus, it might hinder the effectiveness and the organization will fall behind if no action is taken to educate and share information regarding big data to their respective workforce and management.

Induction programs, in-house workshop, and training, as well as inviting experts on each field to talk about the matter can be some means and methods for organization, companies, and institutions to disseminate useful information. It will also help them to deeply understand on how big data can play a major role to find business chances in their industry or market. This will also help in preventing people in the organization to be ignorant and refuse to embrace this new technology.

Explore: By this stage, the organization needs to develop a strategy or set of strategies and road map for big data development. Both strategies and road map will be unique to each and every organization as it will be based on their business needs and challenges. For example, organization or businesses that are in the food industry might not have the same strategies or road map as compared to small–medium food stall.

IBM Institute for Business Value in collaboration with Saïd Business School, University of Oxford (2012) stated that nearly 47% of the respondents reported that formal and ongoing discussion within their organization is about how to use big data to solve important business challenges. Furthermore, another important matter that must be taken into account is to develop quantifiable business case and creating a big data blueprint. This includes gathering of existing data, technology, and skills, as well as a proper plan on when and where to start and align the plan with their organizational business strategy.

Engage: In the engagement stage, organizations will then make interactions and integrate their big data tools and technologies and assess it against their current technologies and skills. It is also at this stage that organizations are to pilot big data initiatives to legalize value and requirements.

IBM Institute for Business Value in collaboration with Saïd Business School, University of Oxford (2012) stated that organization are currently developing proof of concept to validate the requirements associated with implementing big data initiatives as well as to clear the expected returns.

However, organizations in this group are working within a defined and limited scope, which is done to understand and test the technologies and skills required to capitalize on new sources of data.

Execute: At this stage, organization or company should have clear understanding and mind-set in what big data is all about. Therefore, at this stage, deployment of two or more big data initiatives are expected and to continue exploring into advanced analytics. It is at this stage that big data and analytic capabilities are widely used and operated as well as implemented within the organization.

IBM Institute for Business Value in collaboration with Saïd Business School, University of Oxford (2012) also stated that only 6% of the respondents reported that their organizations have implemented two or more big data solutions. The small numbers of organizations in the execute stage is consistent with the implementations that we can see at the marketplace. Regardless, these leading organizations are leveraging big data to transform information assets.

However, with the rapid rate and fast-pace emergence of data adoption, it is most likely that more organization and companies adopt the big data approach.

5.3.2.11 TECHNOPRENEUR AND TECHNOPRENEURSHIP

Technopreneur is the person who produces new forms of organizations, achieving new raw materials and presenting new products and services in order to destroy the existing economic. Also defined as who observe chance and generates an organization to pursue it.

A person is willing to take risks that has the opportunity of profit. Technopreneur defined themselves through their capability to gather and handle knowledge as well to mobilize resources to reach a business and social aims (Susanto, 2016a; 2016b; Almunawar et al., 2015b; 2012a; 2012b).

Another definition of technopreneur is a person who is brave, creative, deviator from established business techniques, and continuously pursues the chance to commercialize new products, technologies, procedures, and arrangement (Susanto, 2016a; 2016b). The technopreneur defined logic from tradition, common tradition from prejudice, prejudice from common sense, and common sense from nonsense while including various ideas from various groups and discipline. The technopreneurs to be skilled in applied creativity that thrives in response to challenge and lack for unconventional solutions. They experience challenges, creative visions solutions, build stories that explain their visions for, and then act to the part of the solutions. They also forge new paths and risk failure, but persistently seek success.

5.3.2.12 ENTREPRENEUR AND ENTREPRENEURSHIP

Entrepreneurship is a way of thinking and acting that is opportunity obsessed, holistic approach and leadership balanced for the purpose of wealth creation. An entrepreneur will actively search for change, respond to it, and exploit it as an opportunity to be one step ahead than others. An innovation is the specific tools of entrepreneur in which they take advantage of it as an opportunity for a particular business or service. Entrepreneurship aim to pursue any opportunity without thinking much of resources available.

The entrepreneurs, creators of new firms are a rare species. In an innovation-driven economy, only 1–2% of the workforce starts a business in any given year. He further stated that innovative entrepreneurs are vital to the competitiveness of the economy and have the ability to create new jobs. However, the benefits of entrepreneurship are to be seen if the business environment is receptive to innovation. Moreover, policy makers also need to plan and prepare for the potential job losses that will be triggered by the "creative destruction" as entrepreneurs strive for increase in productivity.

He also mentioned that entrepreneurs can improve the economic growth by introducing innovative technologies, products, and services. An intense competition between entrepreneurs will give challenges to existing firms to become more competitive. Furthermore, entrepreneurs will be able to create job opportunities both in the short and long run. The entrepreneurial activity also raises the productivity of firms and economies and it can influence structural change by replacing established and sclerotic firms.

Big data can be used by the entrepreneur or organization in recruitment of talent, that is, enterprises will be competing in acquiring and recruiting talent, which is the main task of the human resource department.

For example, when the organizations are in need of new workforce, the managers will convey the task to the human resource department where the department will send the recruitment message through the corporate portal. Then, applicants who are interested in the vacancy or position will be submitting their resume or curriculum vitae. When this is done, human resource will select the applicant's resume and interview appropriate candidate until they find the candidate that they want. A huge amount of data will be gathered and selection process, as well as the recruitment process will be better as companies and organizations are having useful information and data for better decision-making.

Innovation: Education and acquiring knowledge are important features for the economy of all organizations (Susanto, 2016a; 2016b). Individual practice shows that the basic learning and the generation of new concepts play an important role in the business. Knowledge is an important issue in the innovation and integration of new technologies.

The background of education of the supervisors, entrepreneurs, and the owners of the business has been known to be main factor explaining innovation in organizations. Further research shows that the organizations could stimulate their processes of innovation and technological improvement by taking part in value chain by the spillover of knowledge and demands from larger organizations, while others consider that this possibility has been overestimated.

Innovation is also defined as adding something new or upgrading some features or characteristics to an existing product or process. For example, a product or service has been built from zero and has worked well to its current state. However, customers or clients will keep on wanting better product or service; thus, innovation is available to successfully exploit new ways or ideas. Innovation begins with critical thinking and creative ideas. Thus, without creativity, there will be no innovation.

The organization can be innovative in their application of big data in pay–performance for their employees. Pay or salary level is one of the most important aspects in attracting employees or workforce to strive better and do well in their jobs or task. It is also considered by most of the employees as an ultimate goal for their efforts. It is also considered to be an effective means to motivate the employees. With big data, companies can record the daily workload, specific content of the work, and task achievement of each employee, which then can be used for cloud computing processing to make the analysis of these data. Therefore, pay–performance standard wages are to be calculated automatically. Thus, it will improve work efficiency and reduce business investment in human capital.

Creativity: creativity is a process by which a symbolic domain in the culture changed. For example, new songs, new ideas, new machines are all related to creativity. Creativity is the skill to make or bring or produce something new into the existing product or services, whether a new answer to a problem, a new concept, ideas, method or device, or a new inventive object or form.

Creativity is the action that affects from the known to the unknown and seeing things that everyone can see while creating connections that nobody else has made.

Creativity is the ability to create, bring into existence, to invent into a new form, to produce through imaginative skill, and to bring into existence something new. He also stated that creativity is not the ability to create ideas out of nothing, but the ability to generate new ideas, combining, changing, or reapplying existing ideas.

Some creative ideas are astonishing and brilliant, while others are just simple, good practical ideas that no one seems to have thought of yet.

5.3.3 ROLE OF BIG DATA OR MAJOR KEY AREAS

5.3.3.1 COST REDUCTION FROM BIG DATA TECHNOLOGIES

With the availability of big data technologies, businesses and organizations can reduce cost using big data technologies in comparison to traditional database management system. Nowadays, terabyte storage for structured data is relatively cheaply delivered through big data technologies such as Hadoop clusters.

According to the report made by Davenport and Dyché (2013), traditional relational database costs $37,000, while $5000 for database appliance and only $2000 for Hadoop cluster. It shows that application for traditional database management system is relatively high than using big data technology. Although the traditional relational database is reliable and easy to manage, cost of having one can be influenced the decision-making process.

5.3.3.2 TIME REDUCTION FROM BIG DATA

Time reduction is one of the roles and solutions of big data. Based on the study by Davenport and Dyche (2013), it is reported that Macy's merchandize pricing adjustment application provides a typical example of reducing the cycle time for difficult- and large-scale analytical calculations from hours to even days to minutes or seconds. The department store chain has been able to reduce time of optimizing pricing of its 73 million items sale from 27 h to just over 1 h. This helps Macy to reprice items much more frequently to adapt to changing economic conditions in the retail marketplace.

5.3.3.3 UNDERSTANDING AND TARGETING CUSTOMER

Based on Marr (2015), it is acknowledged that one of the roles played by big data is to be aware of the needs and desire of the consumers as well as to discover probable consumers and capable to bring explicit good and services for the consumers. Big data is used to better appreciate consumers and their manners and preferences. Companies are keen to develop their conventional data sets with social media data, browser logs as well as text analytics and sensor data to get an entire picture of their consumers. Its main purpose is to create analytic models.

The company uses big data to recognize the majority precious consumer now and tomorrow. It is prepared without punishing consumer basis of data, yielding tiny answers.

Big data examines a large variety of sources that include planned information such as purchase history, customer relationship management (CRM) information, and intelligence from business associated, as well as shapeless information such as community media. In the case of the airlines, they could include credit card companies, hotels, and other travel industry sources.

He also mentioned that big data analytics also brings formless data into the fold-up, information gleaned from community media feeds, blogs, videos, and other sources. Categorization through this information would help the airline respond a big-picture question that companies have struggled for decades to answer: How do we take care of all our customers like rock stars?

Growing consumer intellect is just one development. As the technology evolves, using big data will speed up three other trends over the coming year.

5.3.3.4 UNDERSTANDING AND OPTIMIZING BUSINESS PROCESSES

Based on Marr (2015), it is stated that big data is also more used to optimize trade process. Retailers are able to optimize their reserve based on predictions generated from community media information, web search trends, and climate forecasts.

One particular business process that is seeing a lot of big data analytics is supply chain or delivery course optimization. Geographic positioning and radio frequency identification sensors are used to track cargo or delivery vehicles and optimize routes by integrating live traffic data and others.

Human resource business processes are also being enhanced using big data analytics. The optimization of talent achievement, as well as the dimension of business customs and staff rendezvous can be acquired using big data tools such as Sociometric Solutions—a company that install sensor into worker's name badges that can distinguish social dynamics in the office. The sensor will then report on how employees go around the office, with whom they talk, and even the tone of voice they use when communicating.

Big data will at last form relations of the value sequence that will help a company make more prepared efficiencies from the present investment. That opinion sphere is formed by data generated in the field, and it is emergent at a speed that is hard to understand. Sensors on a single business airliner produce 20 TB of data an hour. Automobiles are reporting back data composed from aboard sensors and dealer service systems. And the growing tide of radio-frequency-identification-equipped vehicles, crates, and packages is also to be considered.

Unbelievable repositories of information, joint with machine-to-machine communication, are fueling a fresh sign of projecting analytics, services that allow tools such as airplanes to resolve their own repairs schedule, alerting the supply chain to make sure that the required parts arrive at the exact place at the correct time.

Big data is moving from the dominion of information scientists into everyday commerce dealings encounters. In call centers, analytics-infused CRM systems can make evaluation by multiple data sources in actual time to propose offers that a spokesperson can present to a consumer. At the doctor's office, analytics included into a health maintenance app may progress outcomes by presenting the general practitioner with knowledgeable suggestions and next steps to think about in treating a patient (Susanto, 2017b; Susanto and Chen, 2017).

Insurance companies, which have long been information driven, will benefit extensively due to initiation of big data. Industry-specific analytics will help them speed claims processing while reducing costs and spotting possible fraud by use of analytics-backed solutions that can verify whether a claim can be processed automatically or must be flagged for evaluation by a specialist.

5.3.3.5 PERSONAL QUANTIFICATION AND PERFORMANCE OPTIMIZATION

Marr (2015) stated that big data is not just for companies or management but also for all of us in person. He mentioned that we can now gain from the data generated from wearable devices such as smartwatches or smart bracelets. Take the UP band by Jawbone as an example: the armband collects data on our calorie consumption, movement levels, and snooze patterns. While it gives an individual the rich insights, the real value is in analyzing the combined data.

5.3.3.6 IMPROVING HEALTHCARE AND PUBLIC HEALTH

Based on Marr (2015), it is stated that the compute supremacy of big data analytics enables us to decipher whole deoxyribonucleic acid strings in minutes and will let us discover fresh cures and enhance understanding and forecasting disease patterns. Wearable devices such as smartwatches and bracelets can be used to collect information to millions of people and their diseases. The scientific trials of the outlook will not only narrow by little sample sizes but could potentially comprise everybody as well.

Currently, big data techniques are being used to keep an eye on babies in a professional early and sick baby unit. By footage and analyzing every heartbeat and breathing pattern of every baby, the unit was able to develop algorithms that can now predict infections 24 h before any physical symptoms appear. In this way, the team can intercede early and save delicate babies in surroundings where each hour counts.

5.3.3.7 IMPROVING SPORTS PERFORMANCE

Marr (2015) also stated that company such as IBM has developed Slam-Tracker tool for tennis tournaments. We use video analytics to track the performance of every player in a football or baseball game, and sensor technology in sports equipment such as basketballs or golf clubs allows us to get feedback (through smartphones and cloud servers) on our game and how to improve it. Elite sports teams also track athletes outside the sporting environment—using smart technology to track nutrition and sleep, as well as social media conversations to monitor emotional well-being.

5.3.3.8 IMPROVING SCIENCE AND RESEARCH

Marr (2015) also stated that knowledge and study is presently being altered by the new potential big data. For example, CERN, the nuclear physics lab has Large Hadron Collider, the world's major and most dominant atom accelerator. Experiments to open the secrets of our life—how it started and work—produce gigantic amounts of data.

The CERN data center has 65,000 processors to study its 30 PB of information. Nevertheless, it uses the computing powers of thousands of computers disseminated diagonally around 150 data centers globally to examine the data. Such computing powers can be leveraged to change a lot of other areas of discipline and study.

The computing power of big data could also be applied to any set of data, opening up new sources to scientists. Census data and other government-collected data can more easily be accessed and analyzed by researchers to create bigger and better pictures of our health and social sciences.

5.3.3.9 OPTIMIZING MACHINE AND DEVICE PERFORMANCE

Marr (2015) also mentioned in his article that big data analytics assists mechanism and procedure grows to be smarter and further independent. For illustration, big data tools are used to function Google's self-driving automobile. The Toyota Prius is built in with cameras, global positioning system as well as dominant computers and sensors to securely drive on the road exclusive of the interference of individual. We can still use big data tools to optimize the routine of computers and information warehouse.

5.3.3.10 IMPROVING SECURITY AND LAW ENFORCEMENT

Marr (2015) stated that big data is applied heavily in improving security and enabling law enforcement. It has been revealed that National Security Agency in the United States uses big data analytics to foil terrorist plots. Others use big data techniques to detect and prevent cyberattacks. Police forces use big data tools to catch criminals and even predict criminal activity, and credit card companies use big data to detect fraudulent transactions.

5.3.3.11 IMPROVING AND OPTIMIZING CITIES AND COUNTRIES

Marr (2015) also stated that big data can be used to develop a lot of aspects of our city and country. For instance, it allows cities to optimize travel flows based on actual instant travel information as well as social medium and climate data. A number of cities are currently piloting big data analytics with the plan of revolving themselves into smart cities, where the transportation communications and utility processes are all connected up, where a bus would wait for a late train and where traffic signals predict traffic volumes and activate to reduce jams.

In such an instance, Long Beach, California, is using smart water meters to notice illegitimately watering in actual time and have been used to assist some homeowners cut their water practice by as much as 80%. It is very important as the state is going through its worst drought in recorded history and the administrator has enacted the first ever statewide water limitations.

5.3.3.12 FINANCIAL TRADING

Marr (2015) also stated that big data purpose comes from monetary trade. Elevated regularity trade, also known as high-frequency trading, is an area where big data finds a set of use nowadays. Big data algorithms are used to formulate trade decisions. The common of the equity trading now takes place through information algorithms that are more acquired into account signals from social medium network and information website to make buy-and-sell judgment in split seconds.

Computers are planned with difficult algorithms that check markets for a set of customizable setting and look for trade opportunity. The programs can be premeditated to work without human being communication, depending on the desires and requests of the clients.

5.3.3.13 BIG DATA AND ANALYTICS "AS A SERVICE"

Structuring an interior big data section stack with petabytes of storage space, a row of blade servers, and a lineup of information scientists is not contained by the contact or a preferred core capability of every business. In a prior time, spreadsheets were the de facto device and top companion of promotion managers, collecting data from campaigns and digest it into irregular but important insight.

The flood of inputs from social medium and other shapeless information that is a feature of big data does not fit the worksheet form any longer. The quantity, diversity, and speed of information have made it too difficult to study using old-school paraphernalia, and not everybody wishes to become an information scientist.

That is where information and analytics presented as a service will help. Companies of every size can use big data by allocating a group of scientists and assets, getting the knowledge they need without venture further than their core competencies or intriguing on a big flat expenditure.

5.3.3.14 BIG DATA AND MOBILE MEANS NEW BUSINESS PROCESSES

Companies become more data driven, it is only usual that those insights unearth their way into the hands of a group who can set them into the act. Mobility will emphasize the impact of big data on both client aptitude and operational effectiveness by making everything straightaway actionable. Equipped with direct decision-making ability and aptitude on your portable phone, you will be able to put into practice new business processes that will change the scenario of how business is done.

Adding mobility to big data means enabling forefront employees with concurrent insights, when and wherever they need them. Those insights will come from combining data with action—information that is changing on the fly—with data at rest. Mobility also enables immediate data compilation from the field, adding to the pool of information that will drive insights in another part of the arrangement.

Despite the benefit of big data, there have always been problems in analyzing the large-scale data. The problems do not occur suddenly but usually after several years. There are few efficient methods to solve the problem of analyzing large-scale data (Susanto and Chen, 2017). They have stated few methods that are constantly used to increase the performance of data analytics and may be able to analyze the large-scale data in a perfect time such as distributed computing, data condensation, density-based approaches, grid-based approaches, divide and conquer, sampling, and incremental learning.

There are many methods of dimensional reduction. One example of reduction method for reducing the input data volume of data analytic is principal component analysis, and sampling method is constantly used for reducing the data computation of data clustering that is used to speed up the computing time of data analytics.

5.4 CONCLUSION

This chapter discussed technopreneur and innovation through big data. Research on this topic is conducted through focus group and through observation and reflection of journals, books, and other website sources. From the beginning, this chapter explained about the technopreneur, the database management system where it serves as an interface between database and end users. The chapter also explained three important elements which help provide concurrency, security, data integrity, and uniform administration procedures.

Technology has become an important aspect nowadays. Therefore, big data plays an important role in data management that has the capability to help companies or organization to gather information with the ability to store huge amount of data,which technology cannot do due to its limitation. Big data can store an unlimited amount of data where the data sizes from terabytes and petabytes are being produced, handled, and managed with ease. The big data elements and description are also fully explained well in result and discussion section where *Vs* are introduced to be able to describe big data that we face now. Many researchers have different opinions and different aspects to describe big data. Some researchers introduced the minimum of 3 *Vs* and the maximum of 10*Vs*.

It shows that the big data have a huge impact on the society, especially to the technopreneurs, companies, and organizations. The five important aspects of big data are also explained in this chapter because the big data is used effectively in several areas. Big data improves security and troubleshooting, used in customer service and enhance customer satisfaction, improve services and products through social media content, and another two important aspects of big data are that it can be used to detect fraud and risk assessment. Daily routine nowadays became easier due to the existence of big data.

Technopreneur, technopreneurship, entrepreneur, entrepreneurship, and innovation are also covered in this chapter, which are related to the big data. A quick review definition and how these can help the economic growth is also included in this chapter. It also covered the importance of technopreneurship and entrepreneurship by introducing innovative technologies, products, and services. They will be able to create jobs opportunities by having an intense competition between them and raise productivity of firms and economies.

In the last section of this chapter, before the conclusion and recommendation, an additional issue was added to result and discussion. That is, cost reduction from big data technologies, where business and organization can reduce cost by having big data technologies rather than using traditional database management system. Other than this, it is able to reduce time from the long hour to just a few hours. To better understand the customers' behavior, big data is also one of the solutions. It is also used to optimize business processes.

Overall, this chapter explains how important and useful the big data for our daily routine, as it happened in any field, not just for companies or government but also for an individual. It is very important in this century as it has impacts in many areas. Technopreneurs are been created due to the existence of big data. Based on our research, we can now benefit from data generated from wearable devices such as smartwatch. Big data has a huge impact on any business and services and creates more advantage as it helps a lot in problem-solving, including improving healthcare, sport performance, science, and research; optimizing machine and device; improving security and law enforcement; and anything that involve big data in everyday life (Liu et al., 2018; Leu et al., 2017; Leu et al., 2015).

5.5 RECOMMENDATION

5.5.1 CULTIVATING BIG DATA ADOPTION

One of the recommendations is to cultivate and nurture big data adoption into different industries and markets. By doing so, by the end of the day, all industries will be interconnected, interrelated, and interdependent on one and another. By sharing data and information, which can be seen and manipulated in different perspective will ensure that collaboration between different industries can be achieved and it will be efficiency and effectiveness in the long run. It is also beneficial for different groups of stakeholders and caters all relative groups and section of the business.

This big data adoption is also driven by the need to solve business challenges, issues, and obstacles. With the advancement of technological science and breakthrough, it is expected that the world is moving toward embracing new innovation, creativity, and the emergence of a new group of people who are given both abilities into technology and business.

Moreover, due to the changing nature of data, different organizations and businesses are starting to explore the unknown, big data potential benefits. At the same time to extract more value from big data and to choose which path and what to do with the data are presented by big data.

5.5.2 CUSTOMER-CENTERED OR FOCUS OUTCOMES

Businesses and organization are needed to be performance centered, especially to focus on its customers and clients. This can be acted as one of the key roles of technopreneurs in the organization, that is, to create new products or services to cater the potential customers and consumers. Not only that, they are also responsible in upgrading and make advancement into existing products or services. In this way, companies are able to add value to the product and it is important to listen and take into account, the needs and wants of each and every customer or consumer because by the end of the day, it is those customers that are paying and give sales, thus profit to the business.

This can be started off by analyzing customer analytics that enable better services to customers as a result of being able to truly understand customer needs and anticipate future buying behaviors. This can be done in gathering data through survey and feedbacks that can be done online or instantly.

Digitization is also an important factor that creates the surge in big data. It has changed the balance of power between individual and organization. Therefore, organizations are needed to invest in the new technologies and advance analytics to gain better insights and ideas into individual customer interactions preferences. Customers, nowadays, be it end consumers or business-to-business customers, want more than understanding. In order to cultivate effectively, it is important to maintain a good relationship with the customers.

5.5.3 ENTERPRISE-WIDE BIG DATA BLUEPRINT

A road map or blueprint is needed to show the vision, strategy, and requirements for big data within an organization when it is critical to establishing alignment between the needs of business users and the implementation of the road map. It is needed that the road map should be aligned with the business common goals, improving its business objectives.

The blueprint must include identifying the key business challenges to which it will be applied, business process requirements that can define how big data will be used, and the architecture which includes data, tools, and hardware needed to achieve it. Moreover, by developing the road map, it will guide the organization through a different approach to developing and implementing the big data solutions to create sustainable business value.

5.5.4 BUILD ANALYTICS CAPABILITIES ON BUSINESS PRIORITIES

In the world today, a wide variety of analytic tools is presented to organizations and businesses, which have shortage of analytical skill. Thus, big data effectiveness can be at a stake. Therefore, organization needs to invest in acquiring both tools and skills. Furthermore, as a result of these matters, it is expected that there will be new roles, career models for individual with the well-balanced and functional IT skills.

Professional development and career progression of the in-house analysts to further upgrade their knowledge and understanding to the next level are important. It is also the responsibility of the students and undergraduates to build up solid analytical skills.

KEYWORDS

- **database management system**
- **entrepreneur**
- **internet of things**
- **technopreneur**
- **unstructured data**

REFERENCES

Almunawar, M. N.; Anshari, M.; Susanto, H. Crafting Strategies for Sustainability: How Travel Agents Should React in Facing a Disintermediation. *Operational Research* **2013a,** *13*(3), 317–342.

Almunawar, M. N.; Susanto, H.; Anshari, M. A Cultural Transferability on IT Business Application: iReservation System. *J. Hospitality Tourism Technol.* **2013b,** *4*(2), 155–176.

Almunawar, M. N.; Anshari, M.; Susanto, H.; Chen, C. K. Revealing customer behavior on smartphones. *Int. J. Asian Business Inform. Management* **2015,** *6*(2), 33–49.

Almunawar, M. N.; Susanto, H.; Anshari, M.The Impact of Open Source Software on Smartphones Industry.In *Encyclopedia of Information Science and Technology,* 3rd ed.; Khosrow-Pour, M., Ed.; IGI Global, 2015, pp 5767–5776.

Almunawar, M. N.; Anshari, M.; Susanto, H. Adopting Open Source Software in Smartphone Manufacturers' Open Innovation Strategy. In *Encyclopedia of Information Science and Technology, 4th ed.;* Khosrow-Pour, M., Ed.; IGI Global, 2018, pp 7369–7381.

Almunawar, M. N.; Anshari, M.; Susanto, H.; Chen, C. K. How People Choose and Use Their Smartphones. In *Management Strategies and Technology Fluidity in the Asian Business Sector;* Ordóñez de Pablos, P, Ed.; IGI Global, 2018, pp 235–252.

Bendre, M. R.; Thool, V. R. Analytics, Challenges and Applications in Big Data Environment. *J. Manag. Anal.* **2016.** doi:10.1080/23270012.2016.1186578

Chong, D.; Shi, H. Big Data Analytics: Literature Review. *J. Manag. Anal.* **2015.** doi:10.1080/23270012.2015.1082449

Dalton, R. J. The Potential of Big Data for the Cross-National Study of Political Behavior. *Int. J. Soc.* **2016.** doi:10.1080/00207659.2016.1130410

Davenport, T. H.; Dyché, J. Big data in big companies.International Institute for Analytics, 2013, 3.

Einav, L.; Levin, J. The Data Revolution Economic Analysis. *Natl. Bur. Econ. Res.* **2013,** doi:10.3386/w19035.

Gupta, R.; Gupta, S.; Singhal, A. Big Data: Overview. *Int. J. Comput. Trends Technol.* **2014,** *9*(5), 266–268. doi:10.14445/22312803/ijctt-v9p150.

Jenkins, J. C.; Slomczynski, K. M.; Dubrow, J. K. Political Behavior and Big Data, *Int. J. Soc.* **2016,** *46*(1), 1–7.

Knapp, M. M. Big Data. *J. Electron. Resour. Med. Libr.***2013.**doi:10.1080/15424065.2013.84771

Kritikos, A. S. Entrepreneurs and Their Impact on Jobs and Economic Growth. *IZA World Labor.* **2014.** doi:10.15185/izawol.8

Leu, F. Y.; Liu, C. Y.; Liu, J. C.; Jiang, F. C.; Susanto, H. S-PMIPv6: An Intra-LMA Model for IPv6 Mobility. *J. Network Computer Applications* **2015,** *58,* 180–191.

Leu, F. Y.; Ko, C. Y.; Lin, Y. C.; Susanto, H.; Yu, H. C. Fall Detection and Motion Classification by Using Decision Tree on Mobile Phone. In *Smart Sensors Networks;* Xhafa, F., Leu, F.-Y., Hung, L.-L., Eds.; Academic Press, 2017, pp 205–237.

Liu, J. C.; Leu, F. Y.; Lin, G. L.; Susanto, H. An MFCC-Based Text-Independent Speaker Identification System for Access Control.*Concurrency Computation: Practice Experience* **2018,** *30*(2), e4255.

Marr, B. (2015). Big Data for small business for dummies. John Wiley & Sons.

Pak, T. N. The Quest for Innovation and Entrepreneurship in Singapore: Strategies and Challenges. *Glob. Soc. Educ.* **2012.** doi:10.1080/14767724.2012.710121

Shu, H. Big Data Analytics: Six Techniques. *Geo-Spat. Inf. Sci.* **2016.** doi:10.1080/10095020.2016.1182307

Susanto, H. Managing the Role of IT and IS for Suppoting Business Process Reengineering. *J. Systems Information Technol.* **2016a.**

Susanto, H. IT Emerging Technology to Support Organizational Reengineering. **2016b.** https://ssrn.com/abstract=2770318
Susanto, H. Cheminformatics—The Promising Future: Managing Change Of Approach Through ICT Emerging Technology. In *Applied Chemistry and Chemical Engineering, Volume 2: Principles, Methodology, and Evaluation Methods*; Haghi, A. K., Pogliani, L., Balkose, D., Mukbaniani, O. V., Mercader, A. G., Eds.; Apple Academic Press, 2017a, pp 313–332.
Susanto, H. Biochemistry Apps as Enabler of Compound and DNA Computational: Next-Generation Computing Technology. In *Applied Chemistry and Chemical Engineering, Volume 4: Experimental Techniques and Methodical Developments;* Haghi, A. K., Pogliani, L., Castro, E. A., Balköse, D., Mukbaniani, O. V., Chia C. H., Eds.;2017b, Apple Academic Press, pp 181–202.
Susanto, H. Electronic Health System: Sensors Emerging and Intelligent Technology Approach. In *Smart Sensors Networks;* Xhafa, F., Leu, F.-Y., Hung, L.-L., Eds., Academic Press, 2017c, pp. 189–203.
Susanto, H. Smart Mobile Device Emerging Technologies: an Enabler to Health Monitoring System. In *High-Performance Materials and Engineered Chemistry;* Torrens, F., Balköse, D., Thomas, S., Eds; Apple Academic Press, 2018, pp 241–264.
Susanto, H.; Almunawar, M. N. Managing Compliance with an Information Security Management Standard. In *Encyclopedia of Information Science and Technology,* 3rd ed.; Khosrow-Pour, M., Ed.; IGI Global, 2015, pp 1452–1463.
Susanto, H.; Almunawar, M. N. Security and Privacy Issues in Cloud-Based E-Government. In *Cloud Computing Technologies for Connected Government;* Mahmood, Z., Ed.; IGI Global, 2016, pp 292–321.
Susanto, H.; Almunawar, M. N. *Information Security Management Systems: A Novel Framework and Software as a Tool for Compliance with Information Security Standard.* CRC Press, 2018.
Susanto, H.; Chen, C. K. Information and Communication Emerging Technology: Making Sense of Healthcare Innovation. In *Internet of Things and Big Data Technologies for Next Generation Healthcare;* Bhatt, C., Dey, N., Ashour, A. S., Eds.; Cham: Springer, 2017, pp 229–250.
Susanto, H.; Chen, C. K. Macromolecules Visualization through Bioinformatics: an Emerging Tool of Informatics. *Appl. Phys. Chem. Multidiscip. Approaches* **2018a,** 383.
Susanto, H.; Chen, C. K. Informatics Approach and Its Impact for Bioscience: Making Sense of Innovation. *Appl. Phys. Chem. Multidiscip. Approaches* **2018b,** 407.
Susanto, H.; Almunawar, M. N.; Tuan, Y. C. Information Security Management System Standards: a Comparative Study of the Big Five. *Int. J. Electrical Computer Sci.* **2011,** *11*(5), 23–29.
Susanto, H.; Almunawar, M. N.; Leu, F. Y.; Chen, C. K. Android vs iOS or Others? SMD-OS Security Issues: Generation Y Perception. *Int. J. Technol. Diffusion* **2016,** *7*(2), 1–18.
Susanto, H.; Chen, C. K.; Almunawar, M. N. Revealing Big Data Emerging Technology as Enabler of LMS Technologies Transferability. In *Internet of Things and Big Data Analytics Toward Next-Generation Intelligence;* Dey, N., Hassanien, A. E., Bhatt, C., Ashour, A. S., Satapathy, S. C., Eds.; Cham: Springer, 2018, pp 123–145.
Syed, A. R.; Gillela, K.; Venugopal, D. C. The Future Revolution on Big Data. *Int. J. Adv. Res. Comput. Commun.Eng.* **2013,** *2*(6), 2446–2451.

Tsai, C.W.; Lai, C.F.; Chao, H.C.; Athanasios, V. V. Big Data Analytics: a Survey. *J. Big Data* **2016,** *2*(21).

Urbancova, H. Competitive Advantage Achievement Through Innovation and Knowledge. *J. Competitiveness* **2013,** *5*(1), 82–96. doi:10.7441/joc.2013.01.06.

Zang, S.; Ye, M. Human Resource Management in the Era of Big Data. J. *Hum. Resour. Sustain. Stud.* **2015,** *3*(1), 41–45. doi:10.4236/jhrss.2015.31006.

FURTHER READINGS

@. Definitions of Technopreneurship and Technopreneurs.2014. https://entreprenheure.org/2014/05/04/definitions-of-technopreneurship-and-technopreneurs/ (accessed Sep 7, 2016).

Big Data: The 4 Layers Everyone Must Know. (n.d.).http://www.ap-institute.com/big-data-articles/big-data-the-4-layers-everyone-must-know.aspx (accessed Sep 7, 2016).

Big Data: What is Hadoop—An Easy Explanation For Absolutely Anyone. (n.d.).http://www.ap-institute.com/big-data-articles/big-data-what-is-hadoop—an-explanation-for-absolutely-anyone.aspx(accessed Sep 7, 2016).

By better integrating big data analytics into healthcare, the.(n.d.). Big Data: The 25 Eye-Opening Facts Everyone Should Know.http://www.ap-institute.com/big-data-articles/big-data-the-eye-opening-facts-everyone-should-know.aspx(accessed Sep 7, 2016).

How is Big Data used in Practice? 10 Use Cases Everyone Must Read. (n.d.).http://www.ap-institute.com/big-data-articles/how-is-big-data-used-in-practice-10-use-cases-everyone-should-read.aspx(accessed Sep 7, 2016).

insideBIGDATA. Big Data Volume, Variety, Velocity and Veracity. [Online] http://insidebigdata.com/2013/09/12/beyond-volume-variety-velocity-issue-big-data-veracity/ (accessed Sep 15, 2016).

Ways Big Data Will Transform Business. (n.d.).http://www.csc.com/big_data/publications/89362/96477-4 ways big data will_transform_business (accessed Sep 11, 2016).

What is Technopreneurship? 2013. https://rhannieannmay.wordpress.com/2013/03/15/what-is-technopreneurship-2/ (accessed Sep 7, 2016).

Why The 3V's Are Not Sufficient To Describe Big Data [Online] https://datafloq.com/read/3vs-sufficient-describe-big-data/166 (accessed Sep 15, 2016).

CHAPTER 6

ENABLING ONLINE EDUCATION: AN ACADEMIC ASPECT

HERU SUSANTO[1,2,*], FANG-YIE LEU[3], and CHIN KANG CHEN[4]

[1]*The Indonesian Institute of Sciences, Jakarta, Indonesia, E-mail: heru.susanto@lipi.go.id*

[2]*Tunghai University, Taichung, Taiwan*

[3]*Tunghai University, Taichung, Taiwan, E-mail: leufy@thu.edu.tw*

[4]*University of Brunei, Bandar Seri Begawan, Brunei Darussalam, E-mail: chinkang.chen@ubd.edu.bn*

[*]*Corresponding author. E-mail: heru.susanto@lipi.go.id*

ABSTRACT

In today's modern society, information system (IS) and big data can be found everywhere. It is so common that higher learning institutions have also implemented them in the academic field. Many improvements in the education system have been made by using IS and big data. They help increase the productivity of stakeholders in the academic world and their level of efficiency as well. This chapter will discuss on how IS and big data is applied in academic. It will focus on how big data transformed the education system, its importance, the application of it in the academic field, and the challenges it faces as a whole.

6.1 INTRODUCTION

An information system (IS) is a group of computer tools, which can be used for collecting, storing, or processing data. Institutions depend on these systems to help enable and support their operations, interact with people, manage their workforce, provide services, and many other processes. Everyday, the amount of data collected in the digital tools grows tremendously than the previous years. As the amount of data increases, the use of IS becomes more essential and thus, with every passing year, the price of IS hardwares have been decreasing significantly as well. Furthermore, the services of computer storage have been shifting from hardware to cloud systems. This is because of the rising concern toward environmental impacts of the use of electric power by computer hardware.

Many ISs are mainly platforms to deliver data stored in databases. A database is a collection of interrelated data organized (Zwass, 2016). Databases are store organized data so people can retrieve them from different criteria easily. Databases help support everyday operations and management of academic institutions.

To manage these data, we depend on the IS to create, control, store, distribute, locate, and access this information. However, the traditional computing solutions are not scalable enough to manage such magnitude of data. These large sets of data volume are known as big data. Big data is the result of collecting a large volume of data across many sites. It is important to have the right IS to cater high volume of data collected in a system. Academic institutions can benefit from big data if they know how to manage it accurately.

Big data is often determined according to the "3Vs" which are volume, variety, and velocity (Kotadia, 2016). The 3Vs can also be described as the quantity, diverse types of data, and rate of flow of information going into organizations that exceed the capacity of a traditional computer. The volume aspect of 3Vs is the size of data. It is difficult to determine the limits of big data, so this aspect is very relative in the education field. Variety is the different types or formats of big data. Therefore, this means that big data in academic collects, analyzes, and provides information with different backgrounds to ensure better learning resources for institutions. Lastly, Velocity is the increasing flow of data and the need for hardware in ISs to carry more and more information, and for software to process these data as quickly as possible. Big data ensures that stakeholders in the academic environment can have a quick access to information needed in their educational processes.

Today, there are many big data technologies created to help tackle these issues. They transform how big data can be analyzed and utilized correctly. The concept of big data is also expected to change the way of learning approaches, for example, e-learning, where more interactions of students and lecturers are encouraged. ISs and big data are intertwined as to make the academic system a more progressive institution. This will be discussed further below.

6.2 LITERATURE REVIEW

6.2.1 ACADEMIC AND THE USE OF TECHNOLOGY

Shafique and Mahmood (2010) stated "educational institutions play a key role in providing opportunities for people to learn information society knowledge and skills". Information strategy is the support system for both academic and ISstrategies. It is helpful in terms of identifying types of information required, where to apply the information in order to aid and facilitate the main activities, or important goals of the strategy for the academic institution (Allen, 1995). It is also needed to identify how valid and relevant the critical hypothesis isbehind every academic strategy within the context of changing the environment and perceptions.

Many developments have been made in the academic institutions. This is in terms of implementing ISs in the academic settings. According to Shafique and Mahmood (2010), there are many participants involved in an academic process where each user has their own needs to acquire certain information. These participants are among teaching staff, students, administration officers, researchers, and the general public. To accommodate these large numbers of stakeholders, several ISs are introduced, such as the United Nations Educational, Scientific and Cultural Organization National Education Statistical Information Systems development program and the Education Management Information System to help countries to "systematically organize information related to the management of educational development" (Shafique and Mahmood, 2010).

These systems help to solve the common problems like precision, completeness, and accessibility of information that are usually faced by stakeholders. Other than that, IS can also help in enhancing decision-making that is associated with the performance and development of students in the class. Most importantly, these students' related information can be stored and easily accessed in the system that can be of use again in the long run.

Some examples of ISs that are education-based, namely the student management system, teacher management IS, and the school management system. The student management system keeps details of students, such as their demographic data, class status, and so forth. Each student is linked to the institution with a unique code. Using these information, the system can track the student in any given institution through the unique code, whereas teacher management IS is used for tracking the induction, lecturer's training, and professional development progress. This is to solve the problems with lecturers that have poor qualifications. Lastly, school management system keeps details of school's location, types of school, numbers of classrooms, bathrooms, library, and others (De Silva, 2015).

Using computer programs such as Microsoft Data or Excel also improves the efficiency of data management and data manipulation. It is good to implement the technology in academic because of the use of multimedia, flexibility, and real-time engagement. Multimedia, in this case, demonstrates a more enriched teaching and learning enterprises (Clayton-Pedersen and O'Neill, 2005). Technology also adds flexibility in teachers on how they present new knowledge and to get feedbacks from their students. For example, when students are asked to demonstrate their learning through multimedia presentations in class, their skills and content of knowledge will be enhanced. There can also be real-time engagement as students can explore from outside their homeland. For example, they could have video conferencing with other students in different countries to study and share their new knowledge (Susanto, 2017a; 2017b; 2017c; Susanto and Chen, 2017; 2018a; 2018b).

Although it is good in using technology in the academic context, there are also reasons to not implement it. An example would be like how not everyone is fond of using ISs in managing academic. According to Marcella and Knox (2004), from their research survey, it was found that over 60% of university staff from the universities under their study reported that there were problems in the current computer interfaces and/or systems. The problems that were identified included the lack of expertise in the data manipulation or use of applications and the lack of knowledge on how or where data is "housed" in terms of data sets interrelations.

Moreover, the cost is an issue in implementing technology in the academic environment. On the departmental level, it is acceptable to receive grants to develop new learning technologies. On the university scale as a whole organization on the other hand, it would require for the state or government to provide a certain budget to implement the new technology. The massivescale of technology implementation would require in building state-of-the-art classrooms to allow the use of the latest technology. Students who do not have access to personal computers will also experience problems while using the latest technology, as they would have to use computers in campus to do their assignments and other course-related materials. This will hinder the progress of the students' performance.

6.2.2 IS IN HIGHER LEARNING INSTITUTIONS

The mostimportant volume of information is mainly available in social media sites and media networks; however, the percentage of useful information is reduced as compared to other data that are readily available and provided by education institutions and business organizations. Big data in the academic perspective, throughout the whole learning process is usually collected by a variety of management ISs. They are very effective and responsive in enabling academic experts to create and deliver the knowledge, manage the content of the website software, monitor participation, and also in assessing performance among learners. A learning management system, for example, is a web-based software application used in the academic environment for delivering the knowledge online, which can be considered as part of e-learning.

The term e-learning, that is, online learning, according to (Susanto et al., 2018) is publicly acknowledged to have a background of "access" from the years since the 1980s, but it does not have its beginnings exposed.

E-learning, in general, is ranged inclusive from how emails are used between students and lecturers to having a whole learning class online or web-based. It is a solution that allows accessibility to training that became essential in complementing with the traditional way of teaching (face-to-face teaching). In complementing with traditional learning experiences, lecturers can still teach their lessons physically in the classroom by also incorporating the use of technology from time to time, such as activities that are done online using the Internet, simulations, virtual laboratories, and online testing (Arabasz and Baker, 2003).

Learning management systems (LMS) are usually used for conducting online courses and other aspects of learning in academic, but it also changes the term e-learning in identifying how the mechanisms used in the learning experiences to be delivered (Nichols, 2003). It can also be called as course management systems and virtual learning environments. Susanto (2018) described LMS as a software used to deliver, track, and also manage learning instructions. It also helps in many institutions in carrying out courses over the Internet and in featuring online collaborations. However, according to them, an LMS is not intended to train or develop, but to manage. This is because the architecture built for the system is required for managing and administrating training is not the same as the framework that is needed to instruct and also learn.

These LMSs come with both advantages and disadvantages. When the right learning strategies are implemented, LMS can increase the motivation of students, learning, class participation, feedbacks, and support during the process of learning. Another advantage is that the system supports all kinds of formats, like multimedia, video, and text. Materials of courses can be accessed at any time as they are available online. Materials on the courses are updated and students can see the changes made in the course materials. Educators can also modify the information in accordance with the students' needs.

However, LMS can be more of a course-centered rather than student-centered. This is because LMS does not support various teaching styles. There is no guarantee that learning can be improved. Moreover, some educators have weak computer and information literacy skills and these skills are needed to use LMS successfully in supporting their teaching. The teaching staff must learn how to operate within the environments and develop the critical perspective of their use of the LMS in teaching (as cited by Sharma and Vatta, 2013). Additionally, many teachers also find it difficult in designing and organizing a mixture of learning activities that are appropriate with the students' needs, teaching skills and teaching styles. However, from information cited by Sharma and Vatta (2013), it is also very easy to convert existing poor teaching practices into an LMS.

An example of a learning management system used in higher learning institutions is a platform used in learning that is based on the function and work of a teacher's principles that provide a capacity in education for educators to be able to design their content of teaching flexibly online and also to be able to collaborate projects to experience constant feedback from the students. As cited by Costello (2013), some critics view as a neutral platform in terms of teaching functions and styles. Further examples of LMS are BlackBoard, WebCT, and ToolBox.

With the help of ISs, records of revision histories and operational data are also stored efficiently. These data storage can help in decision support and to prepare institutions from making the wrong decision on certain subjects. Additionally, by having accurate and up-to-date information available, institutions can give a better quality service in education. Relationships with consumers, such as the stakeholders in academic institutions, can also be maintained and improved. For example, with ISs, they can get updates from their institutions through their email addresses. This can improve efficiency, as they no longer have to come to announcement boards to see for updates, any social activities, or timetable changes.

6.3 CHALLENGES OF BIG DATA

There are many challenges faced by organizations and institutions in handling big data. Big data is known to be a group of large data sets and is rich with information that is ready for users to analyze and extract. For many years, big data is still growing massively and until now, leaders are still figuring out how to deal with this particular challenge. This is due to the difficulty to match high demands of big data because the degree of coordination and control is lacking in these areas. Visualization can help in performance analysis and more efficient decision-making; however, the issue here concentrates on the large volume of data as well as extracting all of the details at high speed (sas.com, 2013). Leaders are trying to solve this challenge by investing in good IS infrastructures to meet objectives and needs by managing big data effectively. Some are resorting toward powerful and increased memory processor to overcome the large amount of data.

The challenges do not only focus mainly on the volume of data, but also from variety and velocity. Thus, to manage big data effectively, leaders must also assess these issues. In the case of an academic institution, data volume problems like overbearing amount of data can come from accumulated of old types of data that have been stored for many years plus the additional new types of data. For example, old and new records or personal details of the students those go into the institution. This increases volume in the storage which results in too much of data that can be difficult to analyze. Quality of data will also reduce if the data is not accurate or timely. Institutions need to have a quality information management system to ensure that data is always clean. It is best to address issues on data quality and invest on a system, rather than having problems later in time.

In the case of Variety, the problems can come from difficulty to analyze many types of information from data. These data can be extracted through many sources, such as those from tabular data (database), social media, video downloaded from the Internet, images, audio, financial transaction, emails, documents, and so forth. With velocity, this involves data streams, availability to access and deliver the data, how quick the data is being made, how quick the data is being made and others. Velocity, as previously described, deals with the speed of data streaming. The key issue in velocity is the high requirements of end users that have streamed data over their personal devices (Almunawar, 2018a; 2018b). Data transfers usually do not take much capacity of systems. Transfer rates are limited, but not the requests for them. Therefore, data transfer is the main issue in big data. As for now, the only solution is by shrinking the size of data being sent in these data transfers (Almunawar et al., 2013a; 2013b; 2015a; 2015b).

Shrinking size of data can be exampled by Twitter interactions. Most of the interactions present on Twitter are mainly texts, which can be compressed at high rates easily. In terms of academic, many data available could also be modified in the same case as Twitter, with the fact that they are mainly consisting of texts. Journals, essays, and much paperwork can be compressed to reduce the size of data volume that can then increase the speed of data transfers, consequently increasing the velocity in these 3Vs.

The real issue of big data is how people utilize it and their initiative to look for patterns that can help institutions make better decisions out of it. Yvonne Genovese mentioned the benefits of implementing Pattern-Based Strategy, emphasizing on how the strategy can help institution by identifying solutions. By implementing the strategy, it can allow users to adapt to changes by identifying opportunity and threats. It also helps to balance the diversity of institutional activities such as by defining creativity and collectivity that can enable users to lead and respond to changes of weak and strong signals namely opportunity or threat (Susanto and Almunawar, 2018; 2016; Susanto et al., 2011). The aim of this strategy pattern recognition is to understand different elements that can come from many areas such as activities, events, objects, and information. These elements may form into new patterns that represent an opportunity that can be transformed into innovation or even as a threat that can cause a disruption to the business strategy or operation (Susanto, 2016a; 2016b). She also mentioned that this could also be achieved through other mediums such as from social computing analysis or context-aware computing. By doing this, it can eliminate chances of making shortsighted decisions by discovering useful information and knowledge from big data through the strategy.

Furthermore, big data information that is readily available is not set for analysis. For example, a group of students' performance reports submitted to the Ministry of Education comprise students from different schools, level of education, and so forth. These data cannot be left in this form to be analyzed. Rather, these data are needed to be extracted in order to be suitable for the purpose of an analysis. Only required information from the whole bunch of data sets is needed to be extracted for further analysis and doing this correctly and in a complete manner is a technical challenge. It is common to assume that the big data is providing correct information when in fact this is not always the case. Some people may not input correct information in the data sets, which can lead to data errors. Existing work on data cleaning assumes well-recognized constraints on big data or well-understood structure; however, for many well-known big data domains, these usually do not exist (Agrawal et al., 2012).

6.4 WHY BIG DATA IS IMPORTANT TO BE MANAGED

The last few years, the information management has changed radically with the development and improvement of systems to exploit them fully, resulting in the dramatic transformation of IS (as cited by Marcella and Knox, 2004). Big data management can be defined as "the organization, administration, and the government of large volumes of both structured and unstructured data" (Rouse, 2016). The aim is to ensure high quality and accessibility for applications purposes. In the academic environment, big data is mostly more important in library usage. Academic libraries are involved with big data in preserving data sets and researching data management. Academic librarians, therefore, have a clear role in managing big data to help improve the academic institutions and quality of education.

According to Boston University Libraries (n.d.), data can be an important and expensive output of a scholarly research process, across all disciplines. These data are an important part of research results evaluation. They also help to reconstruct events and processes leading to them. The value of data increases as they become aggregated into collections. When they become more available to be reused in addressing new and challenging research questions, the value is unimaginable. That is why the data must be managed properly. Without proper data management, value of these data can be greatly diminished.

Big data can be managed through data mining and data analytics. This is to make mining information easier to gain more insights concerning student performance and learning approaches (West, 2012). With data mining, it can make student-related works, such as evaluation, research, and accountability manageable to address. For example, instead of analyzing the student's test performance, lecturers can use another approach by finding out the most effective teaching method for each student to improve their performances, whereas for data analytics, lecturers can use it to focus on different ways of learning. For example, with online tools, lecturers can obtain the solutions on a much broader range regarding the student actions, how fast they master key concepts, where they usually get their electronic resources, and many more (West, 2012). By doing this, lecturers can get the right information to flow to all students in order to improve their productivity and success rates.

Moreover, it is also important to manage big data to help with data analytics. According to Rouse (2016), data analytics include the steps required in the inspection of large data sets in order to be able to analyze repetitive models, relationships between data and information, and so forth. It is required to store and calculate data in a favorable time and precise and accurate decisions. The findings from data analytics can be used to improve the efficiency of an organization and customer service, to name a few. The efficiencyof an organization can be improved through data analytics by understanding how the organization is working and the efficiency of its performance. Thus, based on the analysis, the organization can strive to make the workforce more effective. This would then result in the boost of performance.

Additionally, from a research study carried out at Educause, it was found that in more than half of the institutions responded under their research, 69% reported that data analytics was viewed as an important priority for at least some departments or programs, while 28% regarded analytics as a major priority for the whole learning institution (Bichsel, 2012). They also found out that respondents said that data analytics have been increased over the last two years, and will continue to increase in importance in the next few years and in the near future. The respondents also believe that data analytics has a great potential to benefit areas that involve students in an institution.

However, according to Herold (2016) with big data analytics, they are not 100% accurate. Data files that are used in the process of data analytics can contain unreliable information, whether these informationare about individuals or other things. If the initial data is already incorrect, then the algorithms or results of the data analytics are bad and cannot be used by other people. The risk of having incorrect data can be dangerous as more data added to data sets could mean that more complex data analysis models will be used without the process of validating whether the information is accurate or not. When decisions are made based on these inaccurate and flawed data models, individuals can suffer harm by being denied services or other treated inappropriately.

It is also important to manage big data because it helps with the process of decision-making. Every organization realizes that information is very crucial in helping to decide on a certain matter, academic institutions are no exception. Many institutions use various sources of information for planning, analyzing trends, managing performance, and other functions. Consequently, according to Susanto (2016a, 2016b) and Susanto et al. (2018), the value of information is only as good as its point of entry into the system. The reason for this is that when decision makers analyzed different information that can lead to making the wrong decisions, which can affect the whole institution, the blame cannot be put on the information itself because the invalidity usually occurs in the process of data entry into the system. Data entry errors and inefficiency in processing information are examples of causes the prone to data errors that are used to make up an institution's decisions.

6.5 MANAGEMENT OF BIG DATA USING CLOUD

According to Hashem et al. (2014), cloud computing is defined as a model for allowing ubiquitous, convenient, and on demand network access to a number of constructed computing resources. These accessibility ranges from networks, server, storage, services, and application. All of these tasks can be supported and supplied by cloud computing with minimal management effort. The cloud is created to handle different kinds of data that can range from external data, internal data, data that comes from personal sources, such as personal mobile phone, tablet, or personal computer, or those data that comes from workers, partners, or even the environment. Organizations, including institutions, rely on this information to carry out everyday activities because it connects the world inside the institution with the world outside the institution.

Managing big data requires high performing IS due to its volume, variety, and velocity nature. Most people usually note on the size of data, but in some cases it does not only imply on the data volume but also the necessity of scalability. When scalability is considered, solutions like traditional database management systems are already out of context. Cloud computing is a paradigm shift of technology that provides computing services over the Internet. It fits the situation perfectly as it offers benefits such as flexibility in using the computing resources, storage capacity, less management effort, and flexible costs (Fernandez, et al. 2014). It composed of highly optimized virtual data centers that have several software, hardware, and information resources that can be used when it is needed by the institution.

Both cloud computing and big data are interrelated because cloud computing provides the basis for big data environment, such as data analysis, accessibility, storage, and easy distribution of information. One advantage of cloud computing is that it is cost-efficient. This is because the services can be deployed without the need of having physical hardwares.

However, although the cloud provides benefits like storage of data as well as data regulating properties, sometimes the flow of data is limited by boundaries that prevent the right data to flow at the right time and at the right place. IBM has come up with a solution to prevent this issue from recurring again. It is known as IBM Cloudant, a database service platform built to ensure flow of data between the application and its database remains uninterrupted and highly performant (IBM Analytics, n.d.). It also ensures that data accessibility for online or offline premises always in smooth sailing without any limitations. The Cloudant also enable users to be connected through interactions. Another service that the cloudprovides in the Cloudant is a built-in self-service data refinery called Data Works. It filters information analyzed from the data. It is very simple and secure enough to use, even for personal data (IBM, n.d.). Organizations or institutions are the perfect candidates to use the cloud because they usually handle a large volume of data that ranges from personal to nonpersonal ones, for example, records and details of students and lecturers, or private and confidential data, like financial information of the institution, and so forth.

Talia (as cited by Hashem et al., 2014) claimed that cloud-computing infrastructures could act as an effective platform to address the data storage needed to carry out big data analysis. Through the cloud, raw data can easily be converted into valuable information, which can serve many purposes in making important decisions. Several applications are offered to users to help them manage big data effectively. For example, applications like Hadoop, Amazon Web Services Elastic MapReduce, Google's Big Query, and Big Data Suite.

A virtual warehouse in the cloud called DashDB can be used to store large amount of data as well as perform analysis (IBM, n.d.). Because the warehouse is cloud based there is no space limitation. The volume of data in the storage can be expanded and scaled if the users are ever to receive new and greater amounts of data. The cloud can simplify the data and further analyze them to extract more information and resources.

Furthermore, implementing the cloud computing can also help manage big data through virtualization. Hashem et al. (2014) stated that it is a process that enables sharing resources that reduces the necessity of having physical hardwares around to increase computer resource utilization, scalability, and efficiency. Through visualization, data can be viewed analytically through different graphs, which can aid in decision-making processes. It is also efficient in terms of being cost-efficient with the help of Hadoop clusters, virtualization can help reduce cost by bringing big data analytics more accessible through reducing computer and storage hardwares. Big data can also help users to use commodity computing to process distributed queries across multiple data sets and get the set of results in a timely manner (Hashem et al., 2014). Cloud computing can also help realizethese objective through the use of a software called Hadoop, a platform that enables data to be processed and distributed. It also provides a service that automatically scales demand of users for data processing.

Other than Hadoop, MapReduce is also preferable for large processing of big data in the cloud environment. This software allows large amounts of data sets to be processed and stored parallel in the cluster. Tools present in MapReduce called Hive and Pig makes data processing more feasible to process large data sets easily. Hashem et al. (2014) mentioned that the cluster computing provides a good support to manage data growth within the context of big data.

Additionally, different cloud models manage big data differently and emphasizes on the importance of having knowledge regarding the three cloud models such as public, private, and hybrid. For example, the public cloud offers scalability and elasticity, which is based on a pay-per-use model. The private cloud offers strict control of which is based on premises infrastructure. Hybrid cloud on the other hand is the mixture of private and public cloud services with combination between the two. Since private cloud has a greater security control over data storage, problems like accidental or malicious access can be prevented through shared resources. For public cloud, it encourages visualization and shared physical resources for data transfers, storage, and processing. However, although private clouds provide tighter security than the two cloud models, public cloud is more favorable and suitable to handle a large volume of big data tasks because of its flexibility. However, potential concerns such as bandwidth limitations and costs incurred with data transfer may be at risk.

Hashem et al. (2014) also highlighted the potential drawbacks that can result from big data management and selecting cloud-computing methods. The first drawback is lack of data availability; users tend to focus their decisions more on analytical methods, which can be expensive. Moreover, incorrect use of methods or methods that inherent weaknesses can result in making wrong and costly decisions. To address these challenges, a common system called Data Base Management System software (DBMSs) is usually used. It plays an important role in ensuring the easy transition of applications from old enterprise infrastructures to new cloud infrastructure architectures (Hashem et al., 2014). These challenges put pressure on users to focus on settling high big data demands and big data storage, which prompts them to acquire the right technologies which is by implementing cloud computing.

6.6 DISCUSSION

Big data plays a significant role in academic institutions. From the analysis of the literature review, it can be noted that the purpose of big data in academic is mainly in the circle of e-learning. Big data has contributed in the rapid progress of the academic environment in terms of online learning. Many systems have been developed to help organize the use of data in higher learning institutions. LMS, especially, have been constantly improved in order to keep up with the vastly growing number of data in the web world. Many higher learning organizations have invested a generous amount of money to have systems installed in their organizations. These systems are hoped to help them in organizing data that can then help in many organizational processes, especially in decision-making.

For example, decision-making in terms of arranging schedules for the students who are taking many different courses in an institution requires organized information at a high speed to have the task done. This would be such a difficult task to do manually as it will take a longer time to organize everything to fit into one schedule without having clashes. The information needed to analyze these are also large in volume; therefore, a system is needed to increase the efficiency in data entry. By using a particular system such as the student management system mentioned previously, data will be automatically entered and analyzed in the system. The details of institutions can also be kept as records in a system such as the school management system that keeps details of types of school and its locations as noted by De Silva (2015).

This is also the same with lecturers; the system can help in identifying problems like capabilities of lecturers in terms of knowledge of teaching. For example, data in the system can be used to track how good the lecturers are doing in their teaching by looking at the student's progress report. This way, the administration team can think to provide solutions by suggesting different teaching techniques to improve the student's grades. This is stated by West (2012) whose statement confirmed that lecturers have many options to explore which teaching methods are suitable for students in order to stimulate better understanding in class. Thus, both IS and big data do not only help in enhancing productivity in students but also in teaching staffs, which is equally important for the institution.

Other than that, another example of this kind of system is the Universities and Colleges Admission Service (also known as UCAS). This is an example of a system that centralizes all information on different universities in the United Kingdom to be easily accessible for international students. This system also helps in registration processes, such as admission of the students' personal statements and acceptance letters from their respective universities. By having this system, many universities will have more admissions and it would be easier for them to accept or reject an application. This is because accepted students can also have a peace of mind, as their offer letters will certainly be sent to them through emails, and not get lost in the mailbox. They will be notified via a message feature in the system if they have been accepted into a certain university.

Other ISs can also aid in decision-making from the financial perspective. Data volume in big data needs to be managed effectively in order to reduce data redundancy or inconsistency. Conn (2012) stated that the increase volume in storage results in too much data that can be difficult to analyze. The quality of data can be reduced if it is not properly managed. This is because data increases in value after it has been analyzed and correlated. Forexample, financial information of an institution is usually in big chunks of data. When these financial information are organized and managed properly, institutions can find out where to invest their money to gain advantages from rather than not using the money efficiently. The budget of organization can also be planned and managed effectively when having this kind of system installed. Errors in calculations can be reduced and this can cut cost-efficient. Although the system might be expensive to install, it is good to invest in the system rather than having financial problems in the future. This kind of IS could be given authorization for it to be managed only by certain trusted officers so money laundering or similar kinds of activities will not be an issue.

In higher learning institutions, the department or group, which uses big data the most is their libraries. Libraries of higher learning institutions consist of much information from scholarly articles, digitized textbooks and journals, graduate research documents, and so forth. The volume of data in library databases must be constantly updated in order for students to have a more recent insight on their research studies or their current modules. This can help improve students' understanding on certain topics and hence increasing the quality of their education overall. Many libraries have authorized storage of outsourced information. For example, Universiti Brunei Darussalam has EBSCOHost, Emerald, and SAGE, to help their students in doing their academic research. For libraries to have a good management IS, old data in their databases can be compressed for the purpose of storing records. This is to save the percentage of storage available in their database system and to avoid system breakdown or slow system performance. This is the reason why the scalability of databases is an important subject to consider in managing data.

The scalability of the IS is an essential feature of managing big data. As the volume of data increases, the computer system must be capable to manage this high volume of data. With every passing year, data entered in the system will certainly increase. With every increase in the usage of data, it also increases the workload of database performance in terms of processing and managing data. If the system is not scalable, it can cause a lot of issues, such as slow performance or down system problem because the system is not equipped to handle heavy load task. This was confirmed by Fernandez et al. (2014) where the traditional database will be out of big data context as the system is not capable to handle big data requirements. However, investing in scalable IS to handle big data ensures full optimization of productivity because it helps the institution to react to new conditions without facing system failure. As a result to this, it is important for the system to have a robust technology that can automatically recover whenever these kinds of problems occur.

Additionally, in order to be efficient in cost-reduction, institutions need to have systems that are high in scalability, which can meet the needs of both current and future data requirements. If the systems are not scalable, they are usually difficult to be repaired and it may take up days to complete which can result in work progress being hindered or in worse case scenarios, they might have to purchase a new database management system every now and then, which is a waste in terms of finance and time. For example, an institution usually implements a system called geographic information system (GIS) that is loaded with a lot of information regarding module registrations, module timetables, examination results, progress report, and so forth. If the system is not performing well, GIS can experience a breakdown, which can cause difficulty for students and lecturers to view information on the system.

Other than that, distribution of information will be delayed which can create further problems. For example, during module registrations, students will have to compete with one another in the attempt to get into their chosen modules due to the maximum quota that is already assigned to each class. If the system is slow, it can reduce the chances of students being accepted to each module which is very inefficient. Furthermore, the accuracy of the data being delivered is also equally important. This is in terms of allocation of timetables. Students and lecturers depend highly on this information to be on track in order for them to be able to divide and manage their time well. If this information is miscommunicated or there are errors in between, it can disrupt lecturers and students' activities as well as the level of their productivity. Therefore, it is important for an institution to test and do research on the systems they want to implement in order to get better results out of them instead of adding more incapable systems. This shows that scalability is an important aspect of big data because it promotes efficiency in terms of conveying information to respective stakeholders as well as helping the institution to be cost-efficient.

The process of data analytics that was mentioned by Herold (2016) is also an important issue that needs to be addressed. In academic institutions, having inaccurate information is not acceptable as it may lead to unfortunate consequences. Students acquire much information from the Internet and other web-based sources as input for their studies. If there are much incorrect information made available on big data, it can create a haywire in the education system. Mistakes in analyzing data can happen in data analytics of academic institutions. For example, students who are tagged with the wrong student identification number can be entered into the wrong classes and this can result in students taking the wrong examination papers.

The use of cloud in academic has also increased in importance ever since it was first introduced. Many higher learning institutions are now investing in cloud systems. This is because the traditional way of keeping information is now considered as costly and ineffective. A paperless-system is more in favor nowadays as the academic environment has also taken part in improving the status of environmental issues. By using cloud systems, many financial costs are being cut and this also promotes a healthier and greener way of running administrations. In going hand in hand with preserving natural resources of the environment, many institutions get marketed better as it can mean that they are morally ethical than those who do not practice "going green." This can also have an effect in student admissions and ranking systems of universities.

As the volume of data increases, the need for increase in velocity is also there. Therefore, cloud systems can help speed up organizational processes in such a way by using the Internet. Much of the information stored in the clouds can be accessible by many stakeholders in the academic field all at the same time. If institutions do not make use of these cloud systems, other institutions can leave them behind. Institutions, which do not use cloud systems, have to do their processes the traditional way (manually) which can be timely. Competitive advantage can be achieved if institutions have a more updated system that is more scalable and also more cost-efficient. High volumes of data can be analyzed and processed at a higher speed. This can then improve the performance of an institution as a whole.

Cloud systems can be used for both educators and learners, for example, students can do their assignments on the go while using cloud systems. It would not matter if they do not have their personal computers with them, they can still access the same information from the cloud. The same goes to educators or lecturers. Computers that are provided by their universities can have the same files as their personal ones at home, providing they do their work in the cloud. As stated by Hashem et al. (2014), the use of commodity computing can distribute queries across multiple data sets and get results in a timely manner.

Although there are many beneficial impacts that can be gained from big data in academic, many issues have also arisen in the process of implementing it and the systems that come with it. First, it was mentioned by Marcella and Knox (2004), not all stakeholders in academic institutions are in favor of using computers. This can hinder the progress of an institution. Additionally, when there is resistance in the institution, the implementation of a certain system is usually unsuccessful. As a consequence, financial investments made to implement the systems will not be fruitful. On the other hand, if there is no resistance, training is also needed for staff and administrators to acquire the use of newly implemented systems. This can increase the expenses of institutions.

Furthermore, the strategies in choosing the right type of cloud system to be implemented in a certain institution are also a challenge. This is a vital process as implementation and installation of the wrong cloud system can be costly and emphasized on having knowledge on the three cloud models such as public, private, and hybrid. Academic institutions might be more suitable in implementing a private cloud for its administration as it provides a tighter security than the other two cloud models. However, public cloud can also be good as it is more suitable forhandling a large volume of big data tasks.

6.7 CONCLUSION

To conclude the whole discussion, big data in academic is a very familiar topic. It has been present for a very long time and progresses have been made ever since. As the number of big data increases, more ISs have also been developed in managing this huge amount of data available. Big data offers many benefits academically. To summarize, it can improve efficiency, effectiveness, and productivity in academic institutions. All stakeholders in academic institutions can also make use of the available big data. Many academic achievements can be credited to the availability of big data in the academic environment. Systems and application softwares that have helped to analyze big data and convert them into useful information is also another subject that can be appreciated.

Without having big data in academic, many organizational processes in academic institutions would have progressed at a slower rate. This would slow down the development of academic achievements of many institutions around the world. Big data has helped improve the academic institutions by providing much useful information. The richer the content of information, more insights can be taken into account. This is the reason to why big data needed to be mined, processed, analyzed, or compressed into aggregated information so that it can be useful to academic institutions.

Furthermore, the availability of cloud computing systems has taken the storage of data to a whole other level. These systems are able to have been allowed for outdated information to still be kept in storage. This old information can be of help in analyzing future trends in the academic environment and not necessarily be a waste of space or data storage. Therefore, a higher number of data can be stored more efficiently. Current technologies are also constantly improving the managing of big data's 3Vs. The volume, velocity, and the variety of big data will always be something that needs to have improvement from time to time.

ISs that have been implemented in academic institutions have also contributed a lot in the development of academic progress. Without the application of these ISs, many processes in the institutions would have taken more time and used more of the institutions' financial budget. With the increasing amount of big data in every passing year, academic institutions would have to constantly prepare for installation of new databases in case the current ones being used are no longer suitable with the needs of the institution. Scalability of databases or system softwares must always be a priority to consider when installing new ones.

Having proper planning in data management on the other hand can reduce challenges of big data in academic. With the help of data management and data analytics, large data sets can be used to improve academic institutions as a whole. Data used to analyze patterns and trends can be stored and computed to help make accurate decisions for institutions. The efficiencyof academic institutions can be improved through data analytics as performance of the institutions is continuously monitored. Consequently, institutions can be more effective and the performance will consistently be improved.

Every set of big data is considered as having a great significance. Each and every information is or can be important in the process of decision-making. This is another reason why it is vital for academic institutions to ensure that there IS is on point and constantly updated. Many institutions sometimes have technical problems occurring in their systems, resulting in the deterrence of their organizational processes. This affects the productivity of many academic institutions. This problem needs to be tackled quickly as to reduce the chances of institutions to be at a cost disadvantage. Performance hindrance can be expensive if it is not fixed. All ISs used in academic institutions must, therefore, be maintained properly to have a constant excellent performance in its systems.

Moreover, many budgets of academic institutions have been allocated in the investment on ISs to handle and manage big data. Although these investments are very costly, it can be approved that it is best to invest on these than to miss out on opportunities that can be grabbed by these institutions. Opportunities in and for the institutional development can be initiated by analyzing information from big data. Therefore, information is needed to be managed constantly and organized efficiently. Data entry errors to systems need to be reduced in order to improve the productivity of academic institutions. Training for technical staff in handling ISs is vital as they are the ones who would be responsible if there is to be any error found in the system. Like data entry errors, individuals and not the system itself cause much of other blunders that occurred in many ISs.

All stakeholders in academic institutions have benefited from big data, especially lecturers and students. Lecturers can improve on teaching materials and have the ability to select teaching styles, which are best fit and students can get better grades from understanding more of their academicals input. With the help of big data in academic, many undergraduate students are able toget their research done by analyzing web-based data sets. Comparisons of articles and journals have been made easy as many of them have been digitized and stored on the Internet. Cloud computing has helped in changing the way students work on their assignments. Although libraries are still important, many students have also been starting to choose to browse for online materials rather than getting physical books from their institutions' libraries.

All in all, it can be concluded that big data in academic has rapidly improved and will continuously increase the rate of progression in academic institutions. It is hoped that in the future, big data in academic will maintain its importance and value as it is today in the current state of theacademic environment, if not better and in a more advanced state.

6.8 RECOMMENDATION

As mentioned above, big data and IS are both important mechanisms in the decision-making process for the institution as a whole. However, there are also challenges faced by the stakeholders that may divert their objectives and hinder them from meeting their goals. Several approaches can be made to eliminate possibilities of making the same errors or new ones in the future. According to Roe (2013), planning, defining strategies, and maximizing existing or readily available data can avoid making potential mistakes and wrong decisions in terms of analyzing big data. By planning and defining the right strategy, the institution will have a clear objective of what they should focus on instead of wasting their time on analyzing data that can result in the meaningless outcome. He mentioned that by implementing analytics strategy, it will enable the institutions to get to know their stakeholders better by understanding their needs. Because of this, nowadays, more people are moving toward technological advances and are starting to take a huge interest in it. Thus, the need for high performing ISs is increasing. By having the right ISs, big data can be regulated efficiently and interactions between stakeholders will also be at ease.

The concept of the analytic strategy focuses on being strategic that can support in decision-making process like addressing problems and questions arise from it. The process involves by starting with a strategic question, finding or collecting relevant data that can help answer the question, analyzing the data as well as making predictions and gaining insight, illustrating findings that can be both understandable and actionable, and lastly, feeding the findings back into the initial process to address the strategic questions and create new ones (Bichsel, 2012). He mentioned that most institutionsare moving toward this strategy because it is different from a lot of traditional analysis and reporting approaches. What separates this strategy from the old approach are data obtained from the system is more extensive and automatic and the processes used to extract and analyze the data are becoming repeatable (Bichsel, 2012). This means that information obtained from the analyzed data can always be used several times whenever the institution needs it. Therefore, because of this, Bichsel

(2012) claimed that the numbers of departments and programs slowly increases in incorporating data and analysis strategy into their decision making and planning process due to the positive feedback.

However, the institution must also ensure that the strategy is able to cover allimportant aspects such as the overall vision and requirements within the institutional context. When this is done, the institution will be able to predict key challenges that might happen in the future as well as identifying solutions to overcome them. Furthermore, the institution can also discover the process requirements that can help to define how big data can be used including which IS infrastructure, tools, software, and many others before officially implementing the strategy.

Moreover, before planning and implementing the strategy, it is crucial for the institution to know if their objectives can be achieved realistically. This can be in terms of whether the strategy can give value to the institution rather than guiding it in the wrong direction. Roe (2013) suggested that the right way to do this is to utilize and maximize the data that is readily available in the institution. This includes utilizing existing software and information surrounding the institution. By doing this, the institution will be able to expand and analyze the information better. Existing software can be modified according to the needs and requirements of the institution to enhance scalability and effectiveness.

Most importantly, the key to managing big data efficiently is to have the right IS. The institution must be able to keep them informed with new and updated technologies as well as software in order to know which one is more fitting to meet the requirements of their stakeholders. This is to provide better tools to manage the constantly increasing big data and to be able to extract the maximum amount of information as possible in order to gain valuable information from it. Other than that, having the right people with the right set of skills is also equally important. Bichsel (2012), emphasizes the importance of investing in a number of analysts before investing in IS tools. These people can serve as a backbone to support major problems like system errors and down system problems that can happen in the future which can deter the process of data analytics. With their knowledge, systems will be restored back to its initial condition. The analyst can also be of help in defining strategies together and ways of solutions to address them. This will not only promote productivity but will also stimulate pool of ideas to help the institution to perform better. Therefore, surrounding the institution with the right skill sets as well as ISs will not only help in big data management but it will also be a good investment which will certainly reward the institution in many aspects.

For example, there are many institutions, which have benefitted from following these steps. According to Bichsel (2012), he found out that most claimed this strategy has helped them in several departments in the institution in terms of how they utilize the data in several functional areas. From the results obtained, many institutions, which have implemented this strategy, manage to spread positive results to all of its stakeholders. In implementing the analytics strategy, most institutions also introduced analytic programs that encourage the stakeholders to participate in. Students and lecturers, who have participated, claimed that data analytics help to foster their interactions, communication, and decision-making as well as boost up their morale in class. For example, it can help with students' enrollment, management, budgeting and finance, and student progress.

Others mentioned how the analytic programs are very systematic in increasing the likelihood of faculty, staff, and administrators to base their decisions on data rather than on intuition or conventional wisdom (Bichsel, 2012). This instilled a form of culture where the stakeholders will treat big data as high importance. Some leaders often follow their intuition or experience or even their preconceived ideas to influence the decision-making process which can reduce the quality of the outcome. This also wastes time and money if they are to implement the wrong strategy. Thus, it can be suggested that analytics strategy helps the academic community to center their attention on strategically important questions only which help to encourage continuous improvement within the institution and among its stakeholders.

He also found out that the strategy could help in increasing satisfaction among the students, teaching staff, administrators, other faculty staff, and so forth. It helps to reduce political conflict between the stakeholders because decisions are no longer biased but will be based on data that have been extracted and analyzed. This can lead to more factual and concrete outcome rather than wasting time and money in extracting and collecting additional data that does not lead to anything. This made the whole process of decision-making more effective, efficient and improved, which is good for the long-term basis of the institution.

Other than helping in decision-making areas, the strategy can also help in enhancing students' progress and their performance as a whole. The strategy can help in extracting data for grade information, demographics, and student personal details, existing and new student data on effort, and engagement for each subject they take. This help to keep track of their early performance as well as study behaviors until the end. When their performance is decreasing in between the semesters, the person in charge will be notified. Steps and different approaches can be taken to improve this issue by consulting the students regarding their overall performance. This approach can result in high grades in students because it can help them be informed of what went wrong with their study and learning methods and how they should tackle the issue. The good part about this is that the academic community can identify students who need additional help that can make the overall performance of the institution better.

Apart from this, the institution must also include the importance of data access by stakeholders in the strategy. The problem of data access is always highlighted because institution sometimes neglects this aspect which can create more problems among the stakeholders such as miscommunication of information, failure to distribute information, or distributing information to the wrong person. To minimize this error, DataMASTER system can be used to create data that is isolated which can be accessible and centralized, data silo promotes a culture of decision making based on data (Bichsel, 2012). AlthoughBichsel (2012) argued that it is impossible to make an isolated for the whole institution, other measures can be taken to ensure data is accessible to everyone. The institution must first need to have a consistent theme, which is a common language around the data so that data can be centralized to make it more accessible and usable. This will ensure data consistency and can be used by various departments. When data is consistent and centrally accessible, it can serve as a better foundation for different variations of decision-making groups.

KEYWORDS

- **E-learning**
- **hardware**
- **information system**
- **learning management systems**
- **velocity**

REFERENCES

Agrawal, et al. Challenges and Opportunities with Big Data: A White Paper Prepared for the Computing Community Consortium committee of the Computing Research Association, 2012. http://cra.org/ccc/wp-content/uploads/sites/2/2015/05/bigdatawhitepaper.pdf</bib>

Allen, D. Information Systems Strategy Formation in Higher Education Institution, 1995. http://www.informationr.net/ir/1–1/paper3.html</bib>

Almunawar, M. N.; Anshari, M.; Susanto, H. Crafting Strategies for Sustainability: How Travel Agents Should React in Facing a Disintermediation. *Operational Research* **2013a,** *13*(3), 317–342.

Almunawar, M. N.; Susanto, H.; Anshari, M. A Cultural Transferability on IT Business Application: iReservation System. *J. Hospitality Tourism Technol.* **2013b,** *4*(2), 155–176.

Almunawar, M. N.; Anshari, M.; Susanto, H.; Chen, C. K. Revealing Customer Behavior on Smartphones. *Int. J. Asian Business Information Management* **2015a,** *6*(2), 33–49.

Almunawar, M. N.; Susanto, H.; Anshari, M. The Impact of Open Source Software on Smartphones Industry. In *Encyclopedia of Information Science and Technology,* 3rd ed.; Khosrow-Pour, M., Ed.; IGI Global, 2015b, pp 5767–5776.

Almunawar, M. N.; Anshari, M.; Susanto, H.; Chen, C. K. How People Choose and Use Their Smartphones. In *Management Strategies and Technology Fluidity in the Asian Business Sector;* Ordóñez de Pablos, P, Ed.; IGI Global, 2018a, pp 235–252.

Almunawar, M. N.; Anshari, M.; Susanto, H. Adopting Open Source Software in Smartphone Manufacturers' Open Innovation Strategy. In *Encyclopedia of Information Science and Technology, 4th ed.;* Khosrow-Pour, M., Ed.; IGI Global, 2018b, pp 7369–7381.

Arabasz, P.; Baker, M. B. Evolving Campus Support Models for E-Learning Courses. 2003. http://www.educause.edu/ir/library/pdf/ERS0303/ekf0303.pdf</bib>

Bichsel, J. Analytics in Higher Education. 2012. http://net.educause.edu/ir/library/pdf/ers1207/ers1207.pdf</bib>

Boston University Libraries. Importance of Data Management, (n.d.). https://www.bu.edu/datamanagement/background/importance/</bib>

Clayton-Pedersen, A.; O'Neill, N. Educating the Net Generation, 2005. http://net.educause.edu/ir/library/pdf/pub7101.pdf#p128</bib>

Costello, E. Opening Up to Open Source: Looking at How Moodle was Adopted in Higher Education. *Open Learning: J. Open, Distance e-Learning* **2013,** *28*(3), 187–200.

DeSilva, S. The Impact of Education Management Information Systems: The Case of Afghanistan, 2015. http://blogs.worldbank.org/education/impact-education-management-information-systems-case-afghanistan</bib>

Fernandez, et al. Big Data with Cloud Computing: An Insight on the Computing Environment, MapReduce, and Programming Frameworks, 2014. http://sci2s.ugr.es/sites/default/files/ficherosPublicaciones/1810_2014-WIRES-Fernandez_etAl-Big_Data_w_Cloud_Computing.pdf</bib>

Hashem, et al. The Rise of "Big Data" on Cloud Computing: Review and Open Research Issues, 2014. http://www.sciencedirect.com/science/article/pii/S0306437914001288</bib>

Herold, R. 10 Big Data Analytics Privacy Problem, 2016 https://www.secureworldexpo.com/10-big-data-analytics-privacy-problems</bib>

Kotadia, H. 5 Ways Big Data Are Fundamentally Changing the Information System, 2016. http://hkotadia.com/archives/5263</bib>

Marcella, R.; Knox, K. Systems for the Management of Information in a University Context, 2004. http://www.informationr.net/ir/9–2/paper172.html</bib>

Nichols, M. A Theory of E-Learning. *Educational Technol. Society* **2003,** *6*(2), 1–10.

Oblinger, D. G., and Oblinger, J. L. Educating the Next Generation. EDUCAUSE, 2005. http://net.educause.edu/ir/library/pdf/pub7101.pdf</bib>

Roe, D. 5 Recommendations for Developing a Big Data Analytics Strategy, 2013. http://www.cmswire.com/cms/information-management/5-recommendations-for-developing-a-big-data-analytics-strategy-019162.php</bib>

Rouse, M. Big Data Management, 2016. http://searchdatamanagement.techtarget.com/definition/big-data-management</bib>

SAS.com. Five Big Data Challenges, 2013. https://www.sas.com/resources/asset/five-big-data-challenges-article.pdf</bib>

Shafique, F.; Mahmood, K. The Role of Educational Information Systems for Survival in Information Society and the Case of Pakistan, 2010. http://www.academia.edu/1190508/The_role_of_educational_information_systems_for_survival_in_information_society_and_the_case_of_Pakistan</bib>

Sharma, A.; Vatta, S. Role of Learning Management Systems in Education, 2013. https://www.ijarcsse.com/docs/papers/Volume_3/6_June2013/V3I6–0456.pdf

Susanto, H. Managing the Role of IT and IS for Supporting Business Process Reengineering. *J. Systems Information Technol.* **2016a.**

Susanto, H. *IT Emerging Technology to Support Organizational Reengineering.* **2016b**. https://ssrn.com/abstract=2770318

Susanto, H. Cheminformatics—The Promising Future: Managing Change Of Approach Through ICT Emerging Technology. In *Applied Chemistry and Chemical Engineering, Volume 2: Principles, Methodology, and Evaluation Methods*; Haghi, A. K., Pogliani, L., Balkose, D., Mukbaniani, O. V., Mercader, A. G., Eds.; Apple Academic Press, 2017a, pp 313–332.

Susanto, H. Biochemistry Apps as Enabler of Compound and DNA Computational: Next-Generation Computing Technology. In *Applied Chemistry and Chemical Engineering, Volume 4: Experimental Techniques and Methodical Developments;* Haghi, A. K., Pogliani, L., Castro, E. A., Balköse, D., Mukbaniani, O. V., Chia C. H., Eds.;2017b, Apple Academic Press, pp 181–202.

Susanto, H. Electronic Health System: Sensors Emerging and Intelligent Technology Approach. In *Smart Sensors Networks;* Xhafa, F., Leu, F.-Y., Hung, L.-L., Eds., Academic Press, 2017c, pp. 189–203.

Susanto, H. Smart Mobile Device Emerging Technologies: an Enabler to HEALTH Monitoring System. In *High-Performance Materials and Engineered Chemistry;* Torrens, F., Balköse, D., Thomas, S., Eds.; Apple Academic Press, 2018, pp 241–264.

Susanto, H.; Almunawar, M. N. Security and Privacy Issues in Cloud-Based E-Government. In *Cloud Computing Technologies for Connected Government;* Mahmood, Z., Ed.; IGI Global, 2016, pp 292–321.

Susanto, H.; Almunawar, M. N. *Information Security Management Systems: A Novel Framework and Software as a Tool for Compliance with Information Security Standard.* CRC Press, 2018.

Susanto, H.; Chen, C. K. Information and Communication Emerging Technology: Making Sense of Healthcare Innovation. In *Internet of Things and Big Data Technologies for*

Next Generation Healthcare; Bhatt, C., Dey, N., Ashour, A. S., Eds.; Cham: Springer, 2017, pp 229–250.

Susanto, H.; Chen, C. K. Macromolecules Visualization through Bioinformatics: an Emerging Tool of Informatics. *Appl. Phys. Chem. Multidiscip. Approaches* **2018a,** 383.

Susanto, H.; Chen, C. K. Informatics Approach and Its Impact for Bioscience: Making Sense of Innovation. *Appl. Phys. Chem. Multidiscip. Approaches* **2018b,** 407.

Susanto, H.; Almunawar, M. N.; Tuan, Y. C. Information Security Management System Standards: a Comparative Study of the Big Five. *Int. J. Electrical Computer Sci.* **2011,** *11*(5), 23–29.

Susanto, H.; Almunawar, M. N.; Leu, F. Y.; Chen, C. K. Android vsiOS or Others? SMD-OS Security Issues: Generation Y Perception. *Int. J. Technol. Diffusion* **2016,** *7*(2), 1–18.

Susanto, H.; Chen, C. K.; Almunawar, M. N. Revealing Big Data Emerging Technology as Enabler of LMS Technologies Transferability. In *Internet of Things and Big Data Analytics Toward Next-Generation Intelligence;* Dey, N., Hassanien, A. E., Bhatt, C., Ashour, A. S., Satapathy, S. C., Eds.; Cham: Springer, 2018, pp 123–145.

Susanto, H. Managing the Role of IT and IS for Supporting Business Process Reengineering. *J. Systems Information Technol.* **2016a.**

Susanto, H. IT Emerging Technology to Support Organizational Reengineering. **2016b.** https://ssrn.com/abstract=2770318

West, D.M. Big Data for Education: Data Mining, Data Analytics, and Web Dashboards, Sept 4, 2012.https://www.brookings.edu/research/big-data-for-education-data-mining-data-analytics-and-web-dashboards/

Zwass, V. Information System, 2016. https://www.britannica.com/topic/information-system

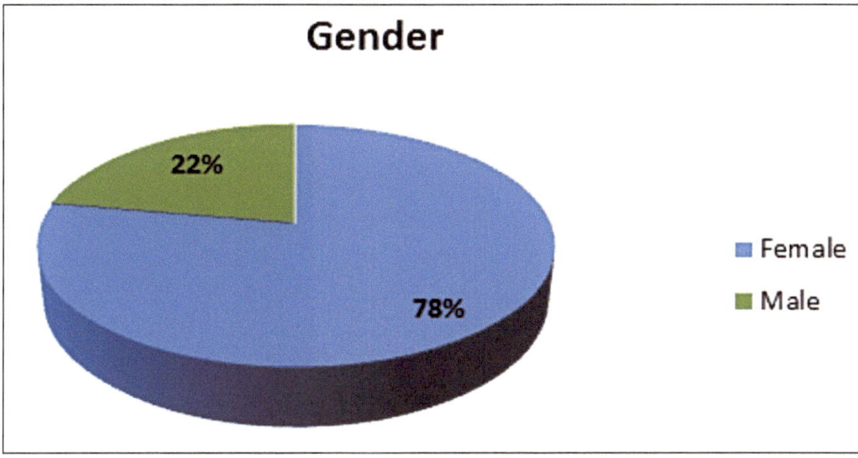

FIGURE 10.1

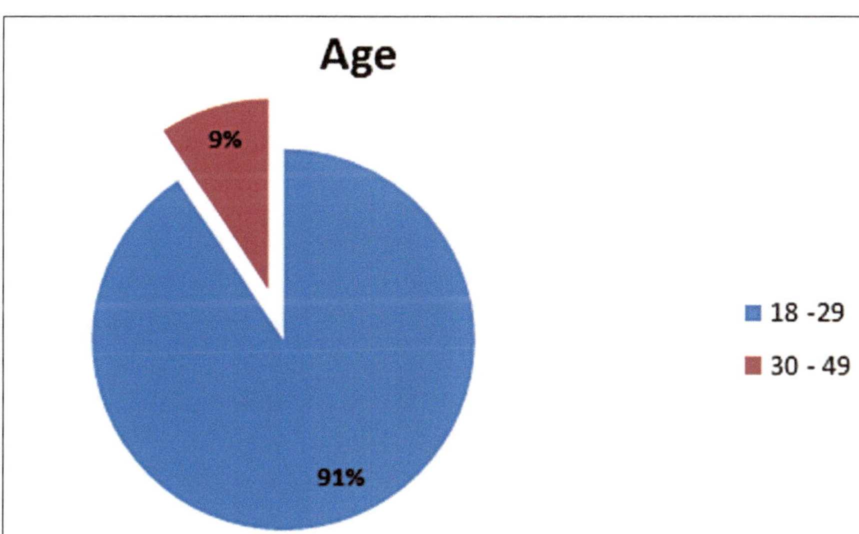

FIGURE 10.2

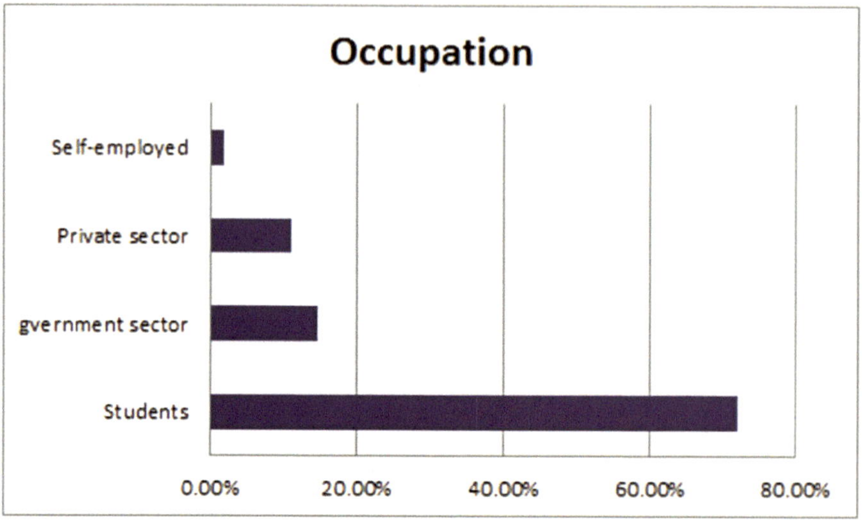

FIGURE 10.3

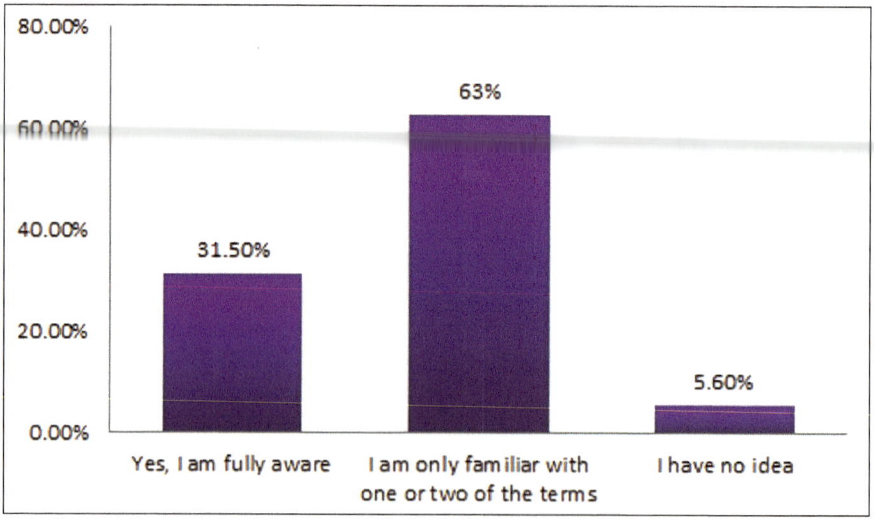

FIGURE 10.4

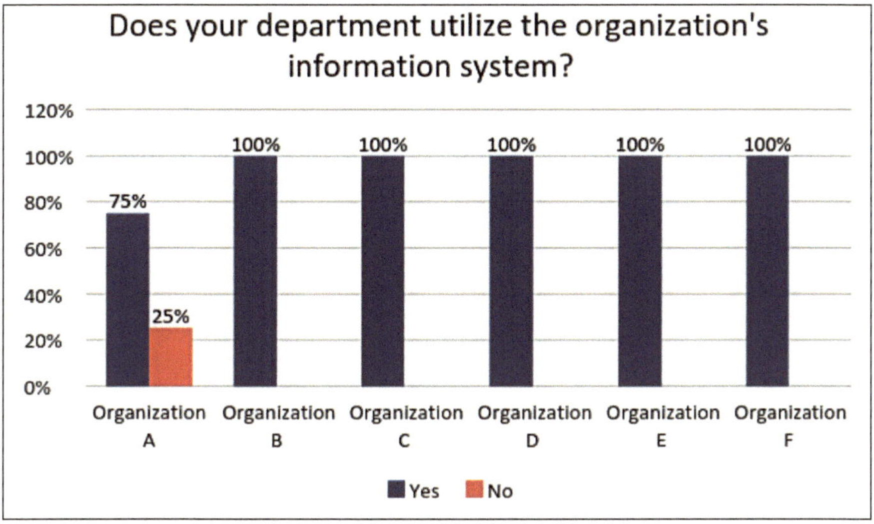

FIGURE 11.1

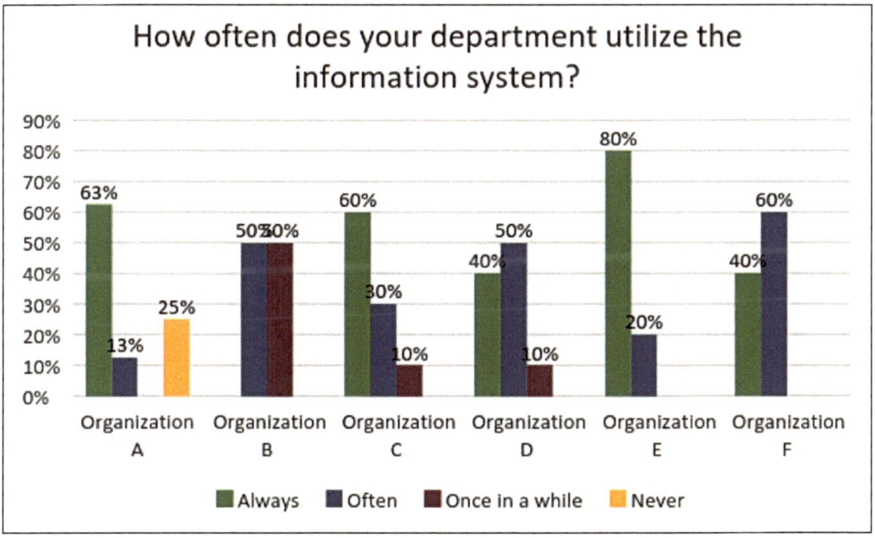

FIGURE 11.2

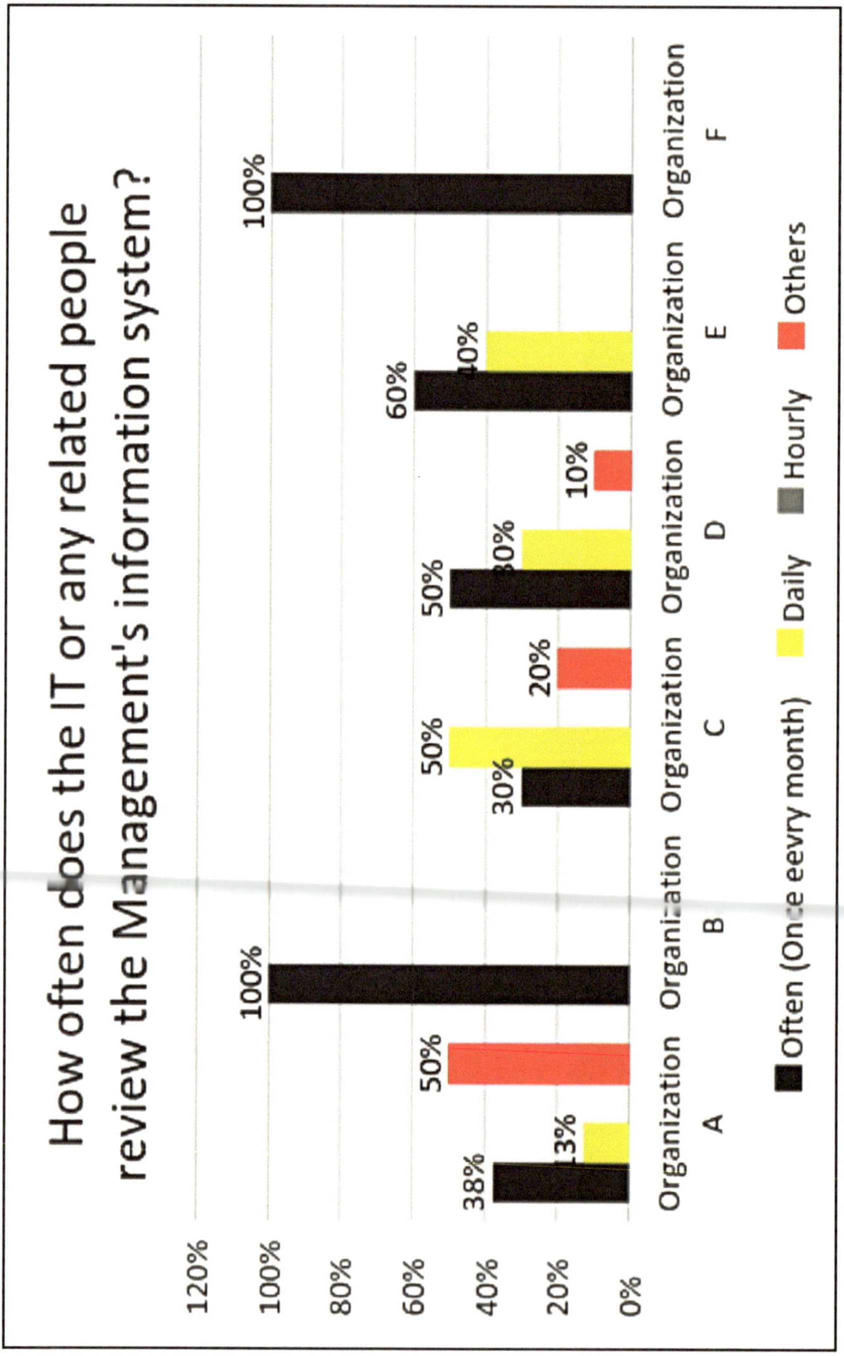
How often does the IT or any related people review the Management's information system?
120%
100%
80%
60%
40%
20%
0%
38%
13%
50%
100%
30%
50%
20%
50%
30%
10%
60%
40%
100%
Organization A
Organization B
Organization C
Organization D
Organization E
Organization F
Often (Once eevry month)
Daily
Hourly
Others

FIGURE 11.3

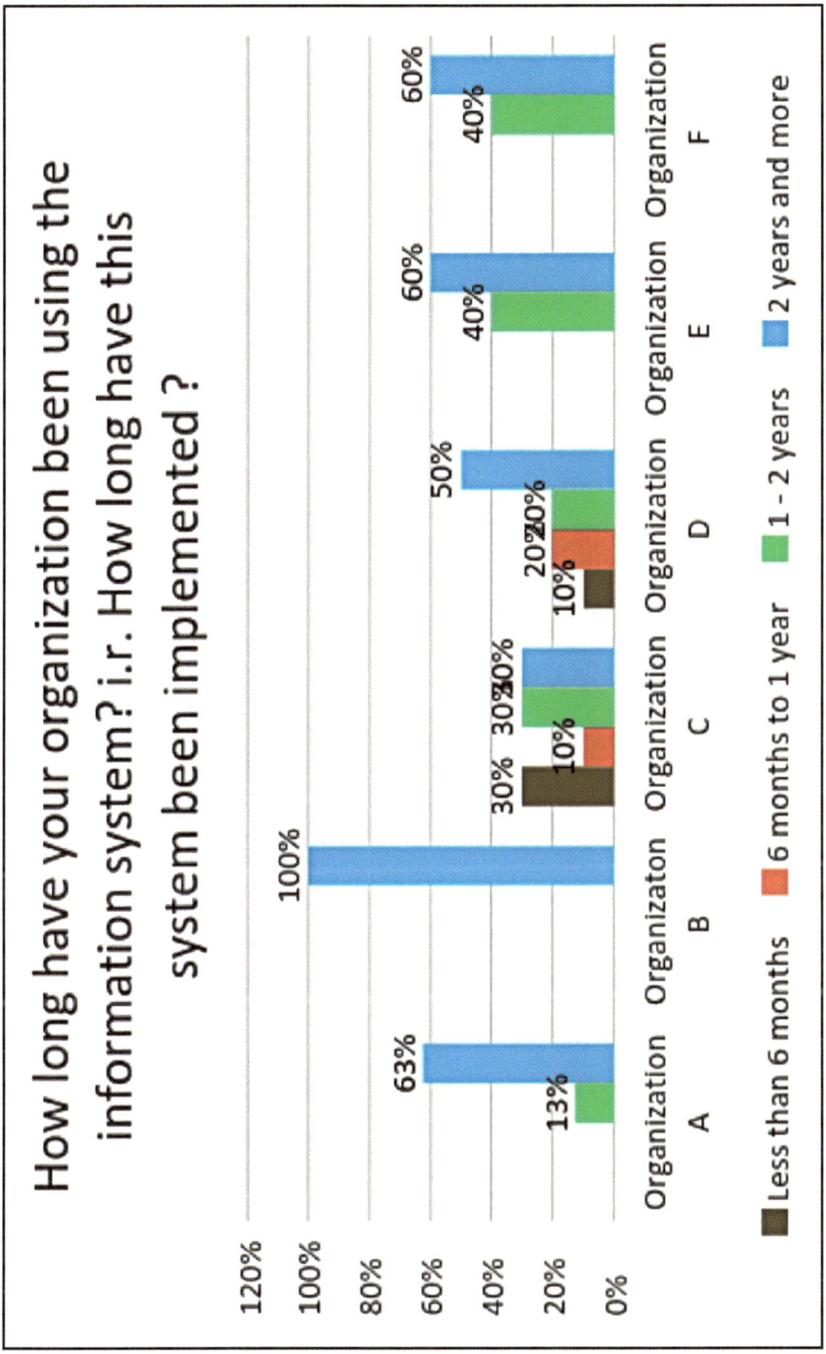
How long have your organization been using the information system? i.r. How long have this system been implemented ?
120%
100%
80%
60%
40%
20%
0%
13%
63%
100%
30%
10%
30%
30%
10%
20%
20%
50%
40%
60%
40%
60%
Organization A
Organizaton B
Organization C
Organization D
Organization E
Organization F
Less than 6 months
6 months to 1 year
1 - 2 years
2 years and more

FIGURE 11.4

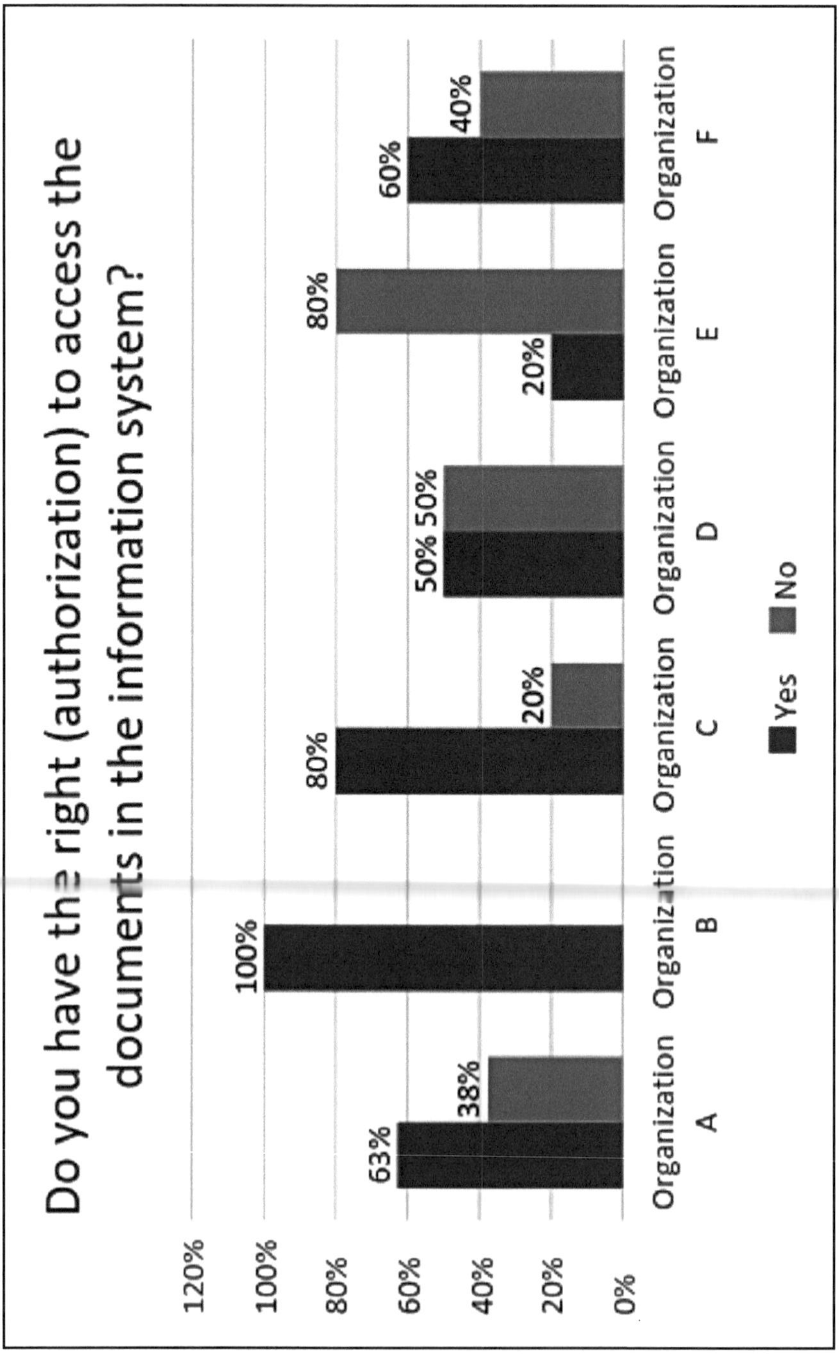
Do you have the right (authorization) to access the documents in the information system?
120%
100%
80%
60%
40%
20%
0%
63%
38%
100%
80%
20%
50% 50%
20%
80%
60%
40%
Organization A
Organization B
Organization C
Organization D
Organization E
Organization F
Yes
No

FIGURE 11.5

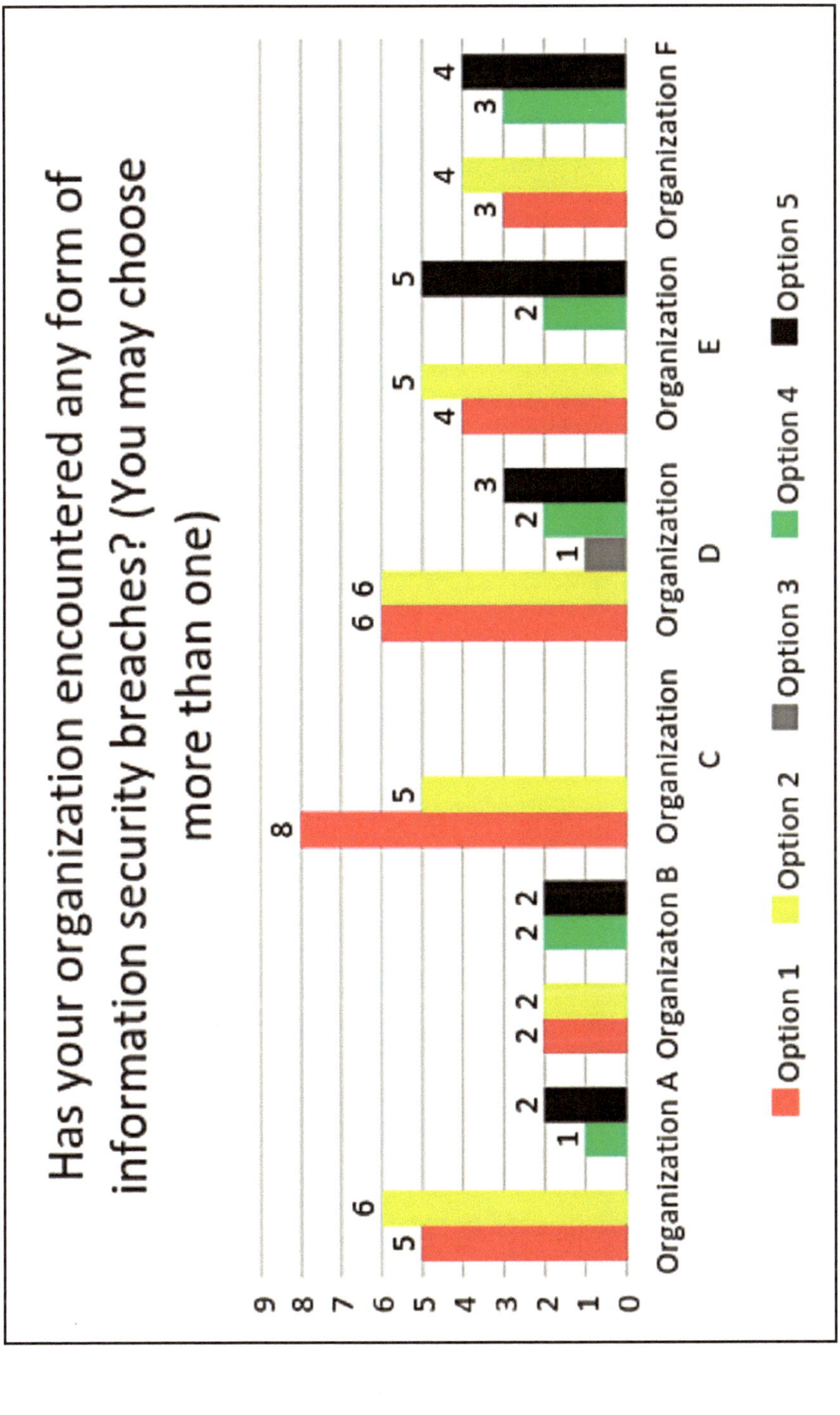
Has your organization encountered any form of information security breaches? (You may choose more than one)
0
1
2
3
4
5
6
7
8
9
5 6
1 2
2 2
2 2
8
5
6 6
1 2 3
4 5
2 5
3 4
3 4
Organization A
Organizaton B
Organization C
Organization D
Organization E
Organization F
Option 1
Option 2
Option 3
Option 4
Option 5

FIGURE 11.6

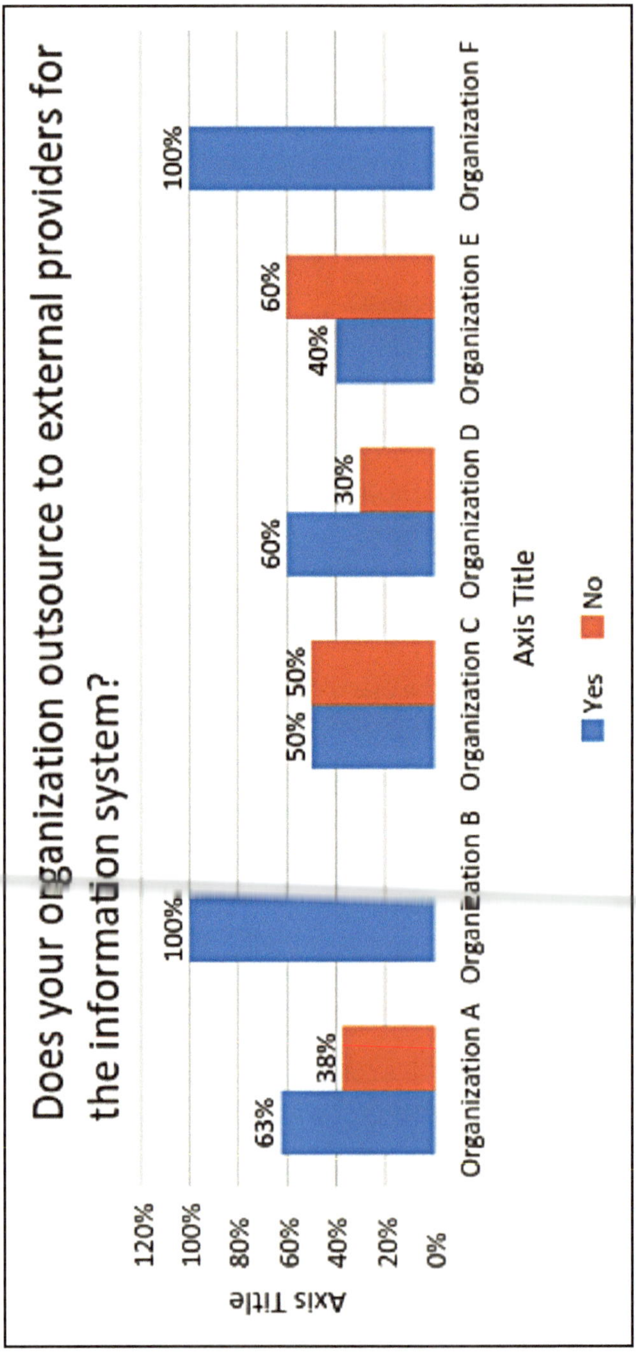
Does your organization outsource to external providers for the information system?
Axis Title
120%
100%
80%
60%
40%
20%
0%
63%
38%
100%
50%
50%
60%
30%
40%
60%
100%
Organization A
Organization B
Organization C
Organization D
Organization E
Organization F
Axis Title
Yes
No

FIGURE 11.7

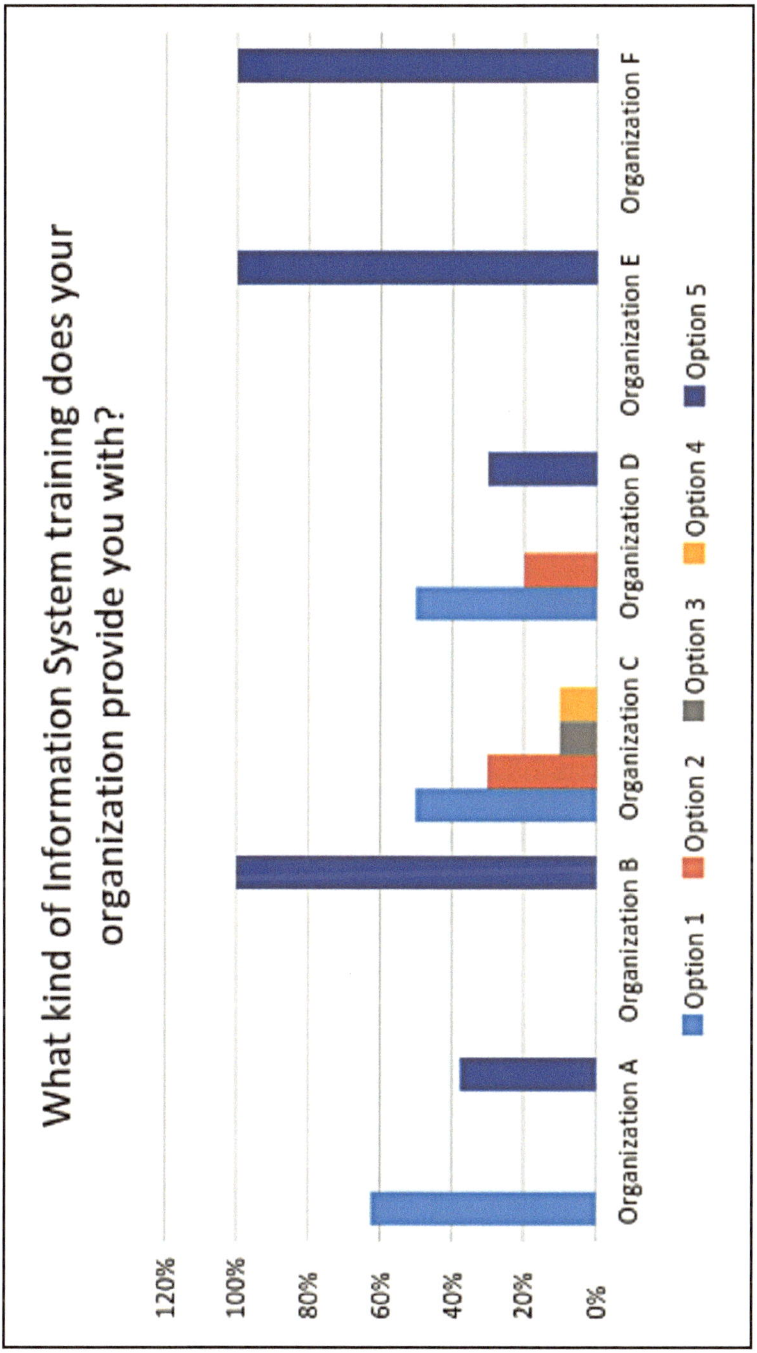
What kind of Information System training does your organization provide you with?
120%
100%
80%
60%
40%
20%
0%
Organization A
Organization B
Organization C
Organization D
Organization E
Organization F
Option 1
Option 2
Option 3
Option 4
Option 5

FIGURE 11.8

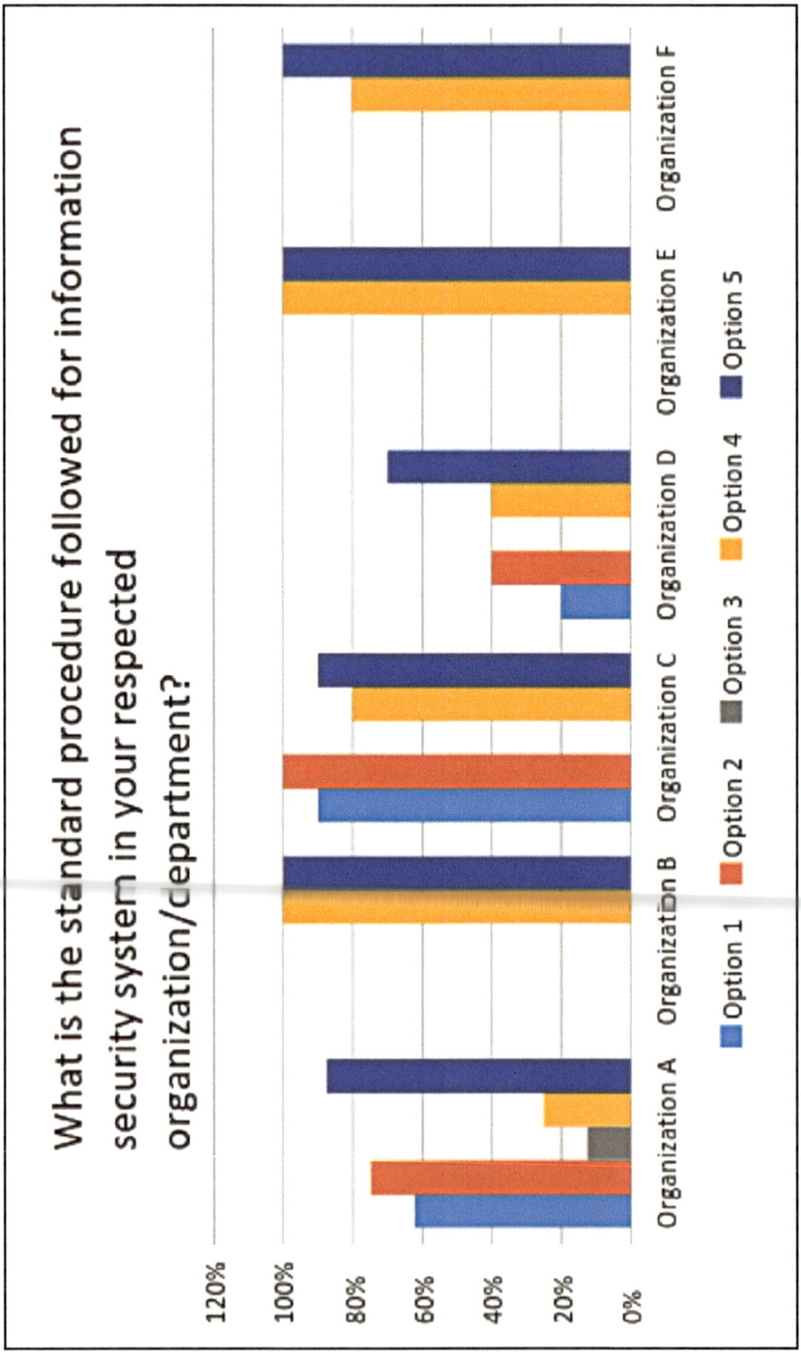
What is the standard procedure followed for information security system in your respected organization/department?
120%
100%
80%
60%
40%
20%
0%
Organization A
Organization B
Organization C
Organization D
Organization E
Organization F
Option 1
Option 2
Option 3
Option 4
Option 5

FIGURE 11.9

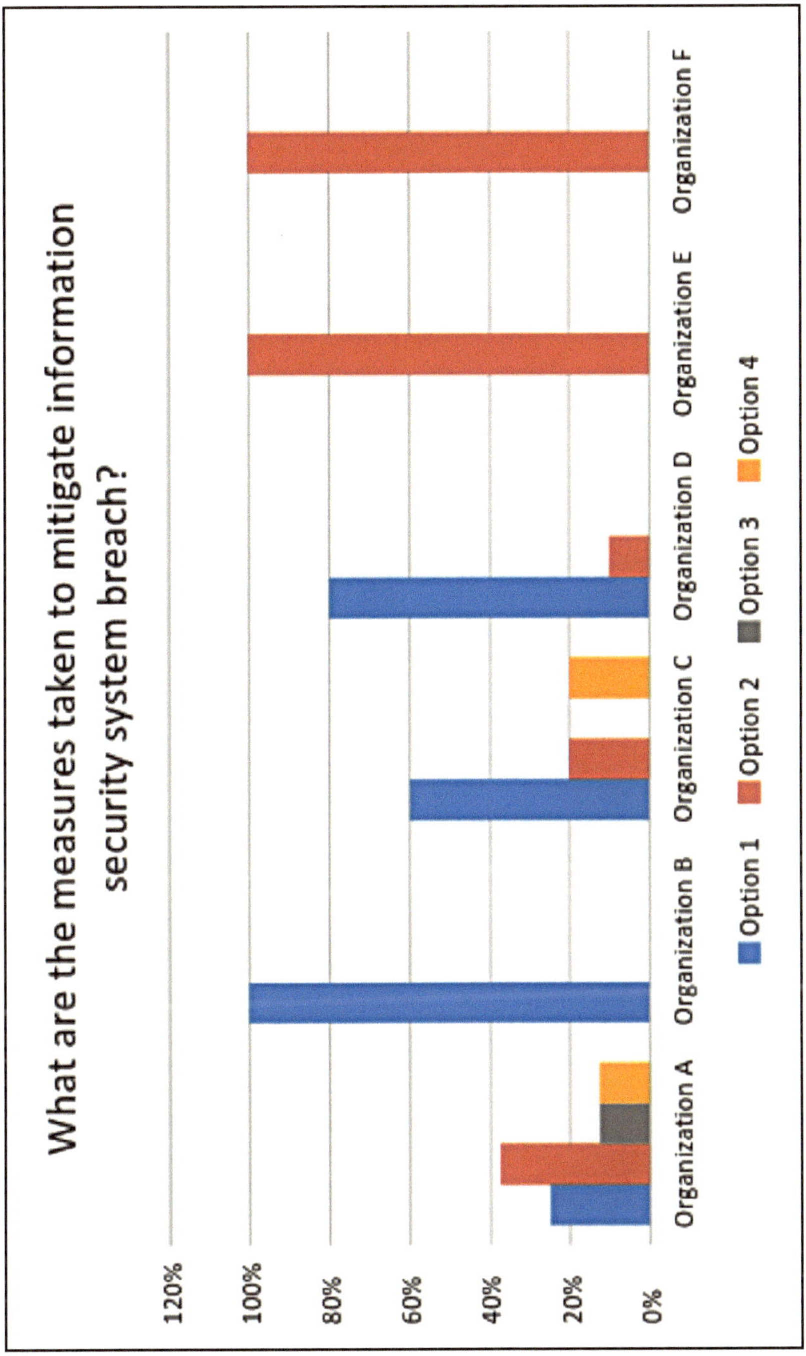
What are the measures taken to mitigate information security system breach?
120%
100%
80%
60%
40%
20%
0%
Organization A
Organization B
Organization C
Organization D
Organization E
Organization F
Option 1
Option 2
Option 3
Option 4

FIGURE 11.10

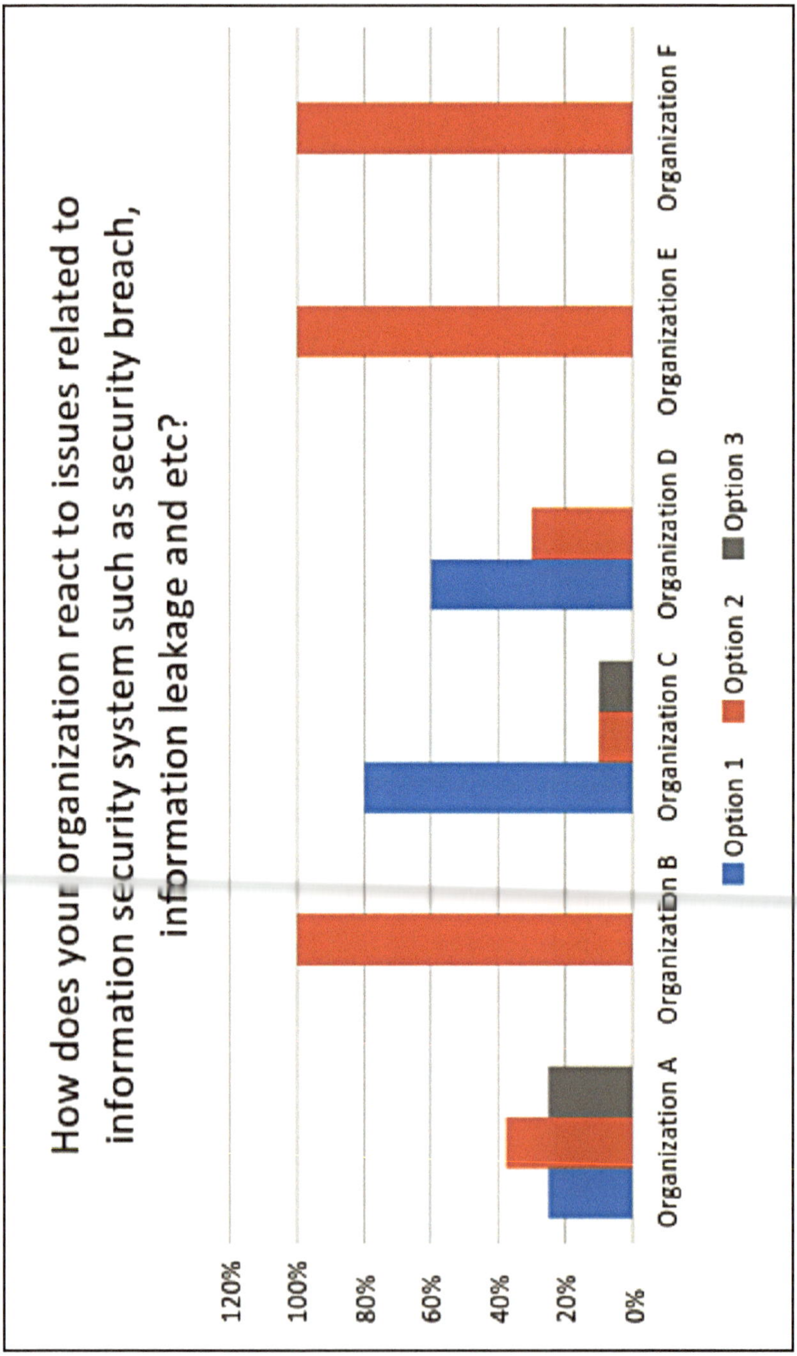

How does your organization react to issues related to information security system such as security breach, information leakage and etc?
120%
100%
80%
60%
40%
20%
0%
Organization A
Organization B
Organization C
Organization D
Organization E
Organization F
Option 1
Option 2
Option 3

FIGURE 11.11

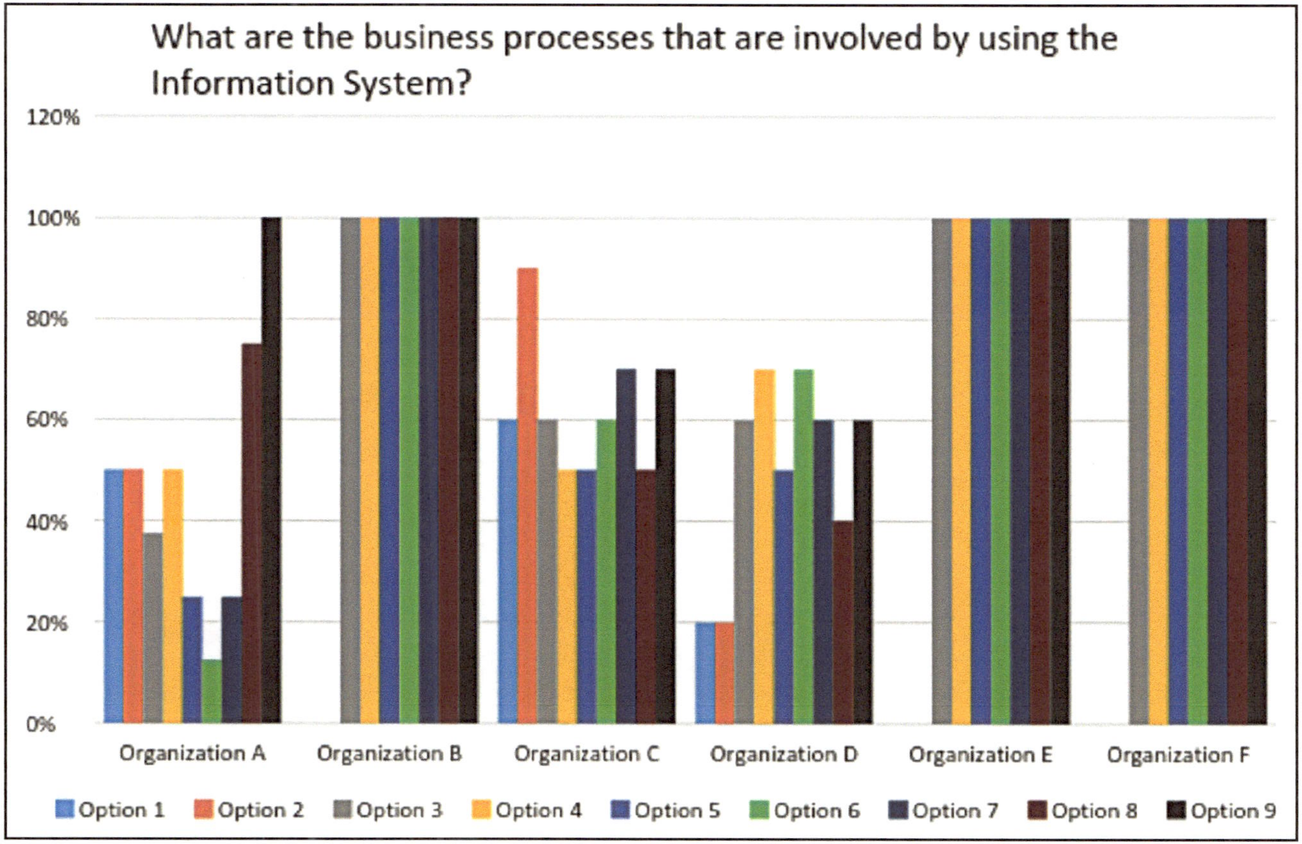
What are the business processes that are involved by using the Information System?
120%
100%
80%
60%
40%
20%
0%
Organization A
Organization B
Organization C
Organization D
Organization E
Organization F
Option 1
Option 2
Option 3
Option 4
Option 5
Option 6
Option 7
Option 8
Option 9

FIGURE 11.12

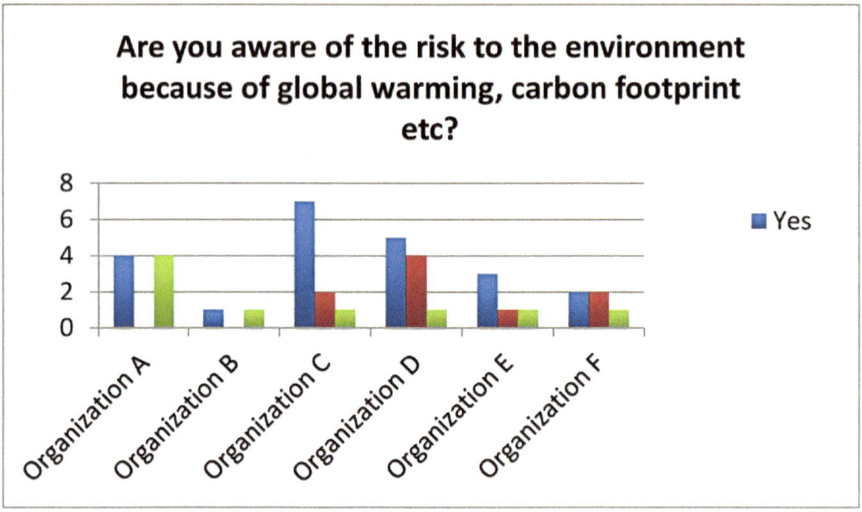

FIGURE 11.13

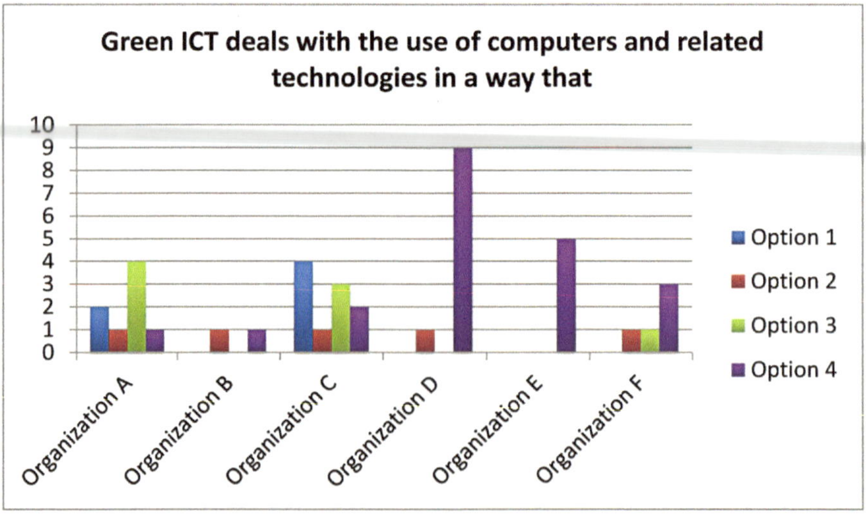

FIGURE 11.14

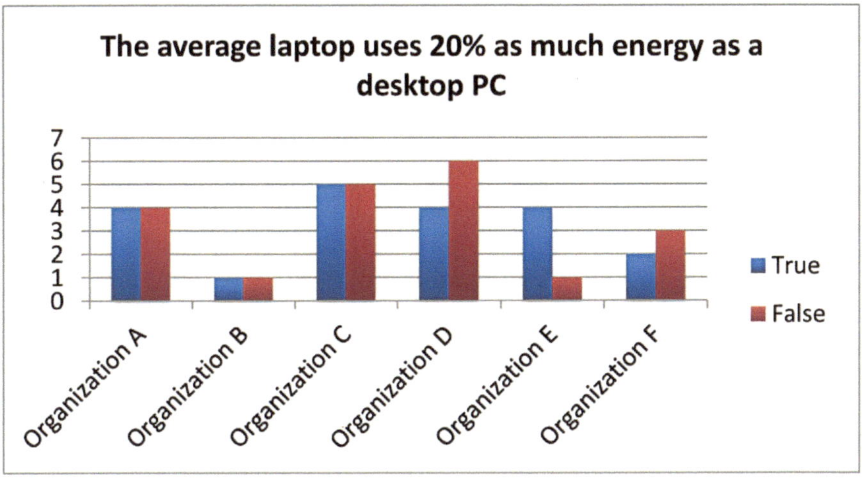

FIGURE 11.15

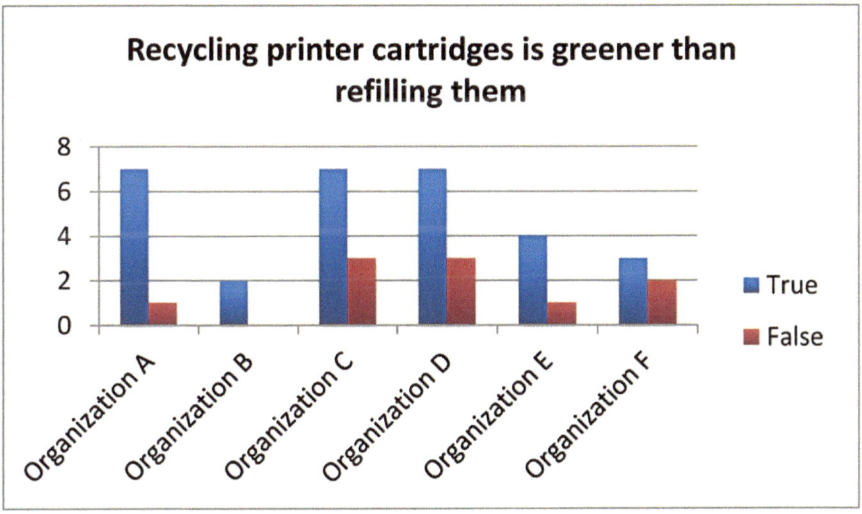

FIGURE 11.16 (See color insert.)

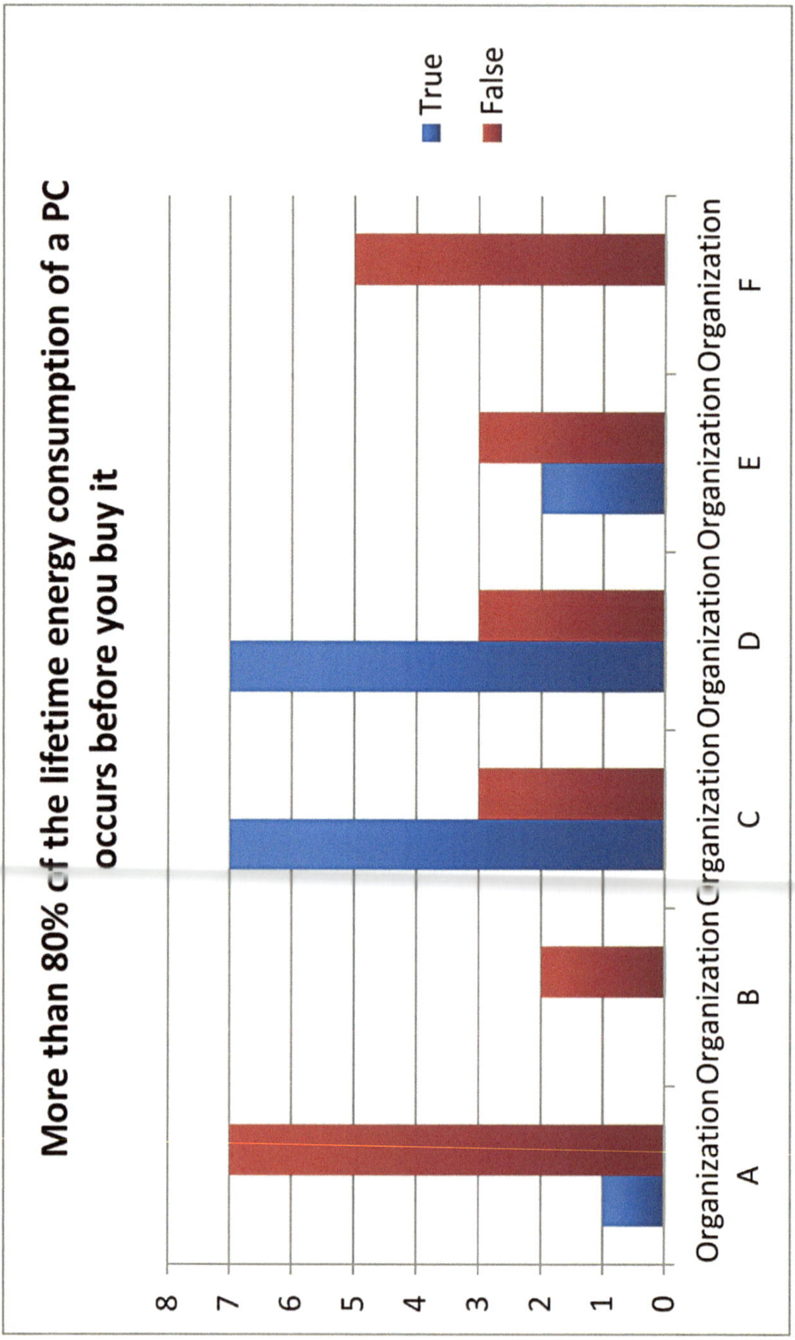

FIGURE 11.17 (See color insert.)

CHAPTER 7

THE EVOLUTION OF LEARNING ANALYTICS THROUGH BIG DATA'S EMERGING TECHNOLOGY

HERU SUSANTO[1,2,*] and CHIN KANG CHEN[3,*]

[1]*The Indonesian Institute of Sciences, Jakarta, Indonesia*

[2]*Tunghai University, Taichung, Taiwan*

[3]*University of Brunei, Bandar Seri Begawan, Brunei Darussalam*

[*]*Corresponding author. E-mail: heru.susanto@lipi.go.id;chinkang.chen@ubd.edu.bn*

ABSTRACT

The application of learning analytics has been growing rapidly in educational institutions in which the learning process is being done through online and hybrid settings. It has become more standardized way of a learning process for educational institutions. The data of the specific behavior of the student will be added into the student-related information. Several main sources used for learning analytics such as virtual learning environment (VLE), student information system (SIS), learning management system (LMS), and library system are apply. The system not only assists in online learning but also creates opportunities for the working group to take up the courses due to its flexibility. This means that it can be seen that technology has the capability to contribute to teaching and learning in the near future. In terms of library system, most of the higher educational institutions in Brunei have their own library system so that it will be easier for the students to make full use of the library resources and provide the students with greater learning experience.

7.1 INTRODUCTION

In this knowledge based era, there are growing amount of data that can be generated rapidly and there is a potential in making full use of the data for learning purposes. The development of learning analytics can actually help determining the performance as the students in schools are analyzed through their skills and knowledge. Additionally, with learning analytics, an educator can access various sources and use relevant information that can be used for teaching materials. This can actually enhance the teaching process as there are more relevant inputs and can also improve student learning experience. The education institution is now acknowledging the development and the use of learning analytics as it is the best education application in enhancing the learning approaches that best suited for the 21st century of education system. The changes in education system are too drastic. This is because the way it is being introduced or put forward is actually making the educational system difficult to cope with it as it is involved with complex processes. The purpose of this chapter is to explore how learning analytics works, how big data contributes to learning analytics, and the challenges of using learning analytics.

7.2 LITERATURE REVIEW

7.2.1 DEFINITION OF THE TERM "BIG DATA"?

Frequently, big data is known as a singular entity when it is actually much more a capability to get information and insight rather than a thing (Sonka, 2016). There is still no exact definition for big data because it is still developing. There are numerous attempts to define what is big data. Big data can be defined as the vast amount of data sets that could be analyzed computationally to uncover the patterns, trends, and associations, particularly it has the connection with human behavior and interactions (Scapin, 2015). According to National Science Foundation, the word big data can be referred as something that is large, different, complicated, longitudinal, and distributed datasets come from instruments, sensors, internet transactions, emails, videos, and other digital sources that can be accessed today and in the future (2012).

7.2.2 CHARACTERISTICS OF BIG DATA?

To understand further on what big data is all about, we need to look into its characteristics. The characteristics of big data can be described using the combination of six Vs which are volume, velocity, value, veracity, variety, and valence.

7.2.2.1 VOLUME

Volume is concerned with how big is the size of the data that is being collected. In the era of big data, the size of the data has reached to terabytes and petabytes turning into metric in order to analyze the data sets. The volume of the data is growing exponentially at an uncontrollable rate. The amount of data being stored in today's time is exploding. In 2000, about 800,000 PB of data were stored in the world. A majority of the data produced currently are not yet analyzed. It is expected that the number of data created by users will reach to 35 ZB by 2020. Basically, the social media such as Twitter alone produce more 7 TB of data every single day, Facebook produces about 10 TB of data, and some organizations produces terabytes of data for every hour of the day of the year (Zikopoulos et al., 2012).

7.2.2.2 VELOCITY

Velocity is concerned with how fast is the speed or the rate of the data created, stored, analyzed, and visualized. The speed or rate of data arrived basically depends on the size of the data. In this case, the smaller is the size of the data, the faster the data will arrive, and the bigger is the size of the data, the slower the data will arrive. This is true nature of data that the size of data determines the outcome of how fast the data will arrive. The improvement on advanced technology has offered people with a high-speed connection and this resulted in producing data at a very high-speed rate. This means that everyday internet users around the world are generating data as they explore through the internet (Rao et al., 2015).

7.2.2.3 VALUE

Value is concerned with the added value of the characteristic of the collected information that could lead to the intended procedure, action, or predictive examination or hypothesis. The value of the data will highly rely on the processes they symbolize like stochastic which means that it is involved in probability, regular, or random process. By relying on this requirement, the collection of data needs to be stored for a long period of time just in case the data will be used as a reference. This means that the data value has a close connection with volume and variety.

There is a huge possibility of the potential value that the big data can offer. For example, McKinsey pointed out about the initiative that new big data can offer under the health care system in the United States. With the implementation of this initiative, it is estimated that the health care spending could be reduced around $300 billion to $450 billion or even about 12–17% of the $2.6 trillion of the overall cost of healthcare in the United States. Nevertheless, the cost of having low value or poor data is also huge. It is estimated that with the availability of low value or poor data used by the US businesses may result in the businesses to incur about $3.1 trillion a year. This means that collecting data without a proper way is considered to be useless (McNulty, 2014; Susanto and Chen, 2017; 2018a; 2018b).

7.2.2.4 VERACITY

Veracity is concerned with how the quality of data is being stored in big data environment and its understandability. The data generated around the world has distinct qualities. It means that different internet users have different way of using the data. Veracity also can be described as how much trust is being put into the data and to some degree, the outcome of data velocity and variety. The high-speed rate in which data is generated and needed to be processed makes it difficult to continuously cleanse and handle the preprocessing in order to enhance the quality of the data.

7.2.2.5 VARIETY

Variety is concerned with the availability of different type of data that can be used. In other word, variety consists of structured, semi-structured, and unstructured data. Previously, people were only focusing on structure data that can be easily fitted into tables or databases like financial information. People should be aware that about 80% of the world's data is unstructured such as texts, images videos, voices, and others. Nowadays, with the development of big data technology, it enables us to examine and bring data together from different kind of data like messages, social media conversations, photos, videos, and voice recordings (Susanto et al., 2018).

7.2.2.6 VALENCE

Valence is concerned with the measurement of the connectivity. The term "valence" comes from chemistry. It is said that the higher the data are connected, the higher is the valence.

7.2.3 WHAT IS LEARNING ANALYTICS?

Basically, learning analytics highly emphasizes on the measurement, collection, analysis, and reporting of learner's information with the aim of comprehending and maximizing learning process as well as the environments in which it is occurring (Attwell, 2016). Learning analytics can be referred as an area of experiments that develops on ideas from different fields in terms of process mining, business intelligence, data processing, information retrieval, and technology-enhanced learning. It is not solely based on technology or institutional statistics. The core of learning analytics is basically concerned with enhancing individual capabilities that will help individual to achieve their potential and goal in life (Hall, 2016; Susanto, 2017a; 2017b; 2017c).

7.2.4 SWOT ANALYSIS OF LEARNING ANALYTICS

In big data era, learning analytics has the possibilities to offer educators with information and tools that will help them making decision and encourage the teachers to make full use of the in a scope of learning concepts. The incorporation of technology for learning purposes gives further information for analysis for supporting an ongoing cycle for continuous improvement. In this case, every educator should be given access to learning analytics tools and given proper training on how to use these tools properly. This can contribute to create e-proficiency and support the development of the targeted and effective way of learning methods.

This was done through examining the case studies that take place within the time frame of 6 years from 2008 to 2013. By analyzing the information, she identifies the strengths, weaknesses, opportunities, and threats of learning analytics.

7.2.4.1 STRENGTHS

The strengths of learning analytics include the availability of huge amount of educational data, the ability to use powerful, preexisting algorithms, the availability of multiple visualizations for staff and students, increasingly precise models for adaptation and customization for learning processes, and increasing valuable insights for learning strategies and behaviors.

7.2.4.2 WEAKNESSES

The weaknesses of learning analytics include the high possibility of misinterpretation of the data, insufficient quality in terms of the sheer variety of data sources, insufficient significant outcomes obtained from qualitative research, involvement in high level of complexity system, and too much information.

7.2.4.3 OPPORTUNITIES

The opportunities of learning analytics include using open-linked data able to enhance the compatibility across the systems, improving self-reflection, self-awareness, and learning process through intelligent systems, and the feeding of learning analytics outcomes to another system to help making decision.

7.2.4.4 THREATS

The threats of learning analytics include the concern of ethical issues specifically on privacy issues, overanalysis, insufficient generalizability of the outcomes, higher chances for misclassification of patterns, and contradictory findings.

7.2.5 RELATIONSHIP BETWEEN BIG DATA AND LEARNING ANALYTICS

There is a correlation between big data and learning analytics. In a numerous point of view on learning analytics, big data is linked to education (NMC Horizon Report, 2013). Learning analytics has been regarded as an educational application of big data. It basically consists of a branch of statistical analysis that was initially created as a method for businesses to make analysis on certain areas such as commercial activities, identifying the spending trends and making prediction on consumer behavior (NMC Horizon Report, 2014; Susanto, 2016a; 2016b).

In generic terms, big data expects the use of either information or database systems as the central storage facility as it is suitable for storing vast amounts of data and noting down to certain transactions. For instance, the record-keeping framework of the undergraduate students has the ability to store the information of the undergraduate students in terms of their grades for each course they have taken. For institutional researchers, this kind of information is very useful and it can be used to analyze the performance of the students over time. It could be analyzed from one semester to another semester or from 1 year to another year. In the context of big data scenario, the information would be obtained for every undergraduate student's transaction in every course, particularly if the course has been conveyed through electronically on the web. This means that every undergraduate student data entry basically on their course evaluation, discussion board section, blog section, and any activity could be recorded and this is producing a large number of transaction for each of undergraduate students per course. Moreover, this information would be gathered on the basis of real time or close to real time every time it is being executed and after that, it will be further examined to recommend the course of action that needs to be taken (Picciano, 2012).

7.3 DISCUSSION

7.3.1 *HOW LEARNING ANALYTICS WORKS*

The application of learning analytics has been growing rapidly in educational institutions in which the learning process is being done through online and hybrid settings. It has become more standardized way of a learning process for educational institutions. Specific kind of web tracking tools has been widely used in educational institutions in order to record the student behaviors in online courses. The tracking tools not only record basic variables like how much the student spend their time on a topic but also more toward nuanced information that could give proof of student critical thinking, synthesis, and the depth of retention of the information after some time. The data of the specific behavior of the student will be added into the student-related information (NMC Horizon Report, 2014).

In this knowledge-based era, a majority of educational activities such as e-learning, tests, quizzes, and other activities are being done through online. Nowadays, students and educators are increasingly connected to the internet because a various amount of the educational material are available in digital format. This has proven that the availability of data source on the internet for learning purposes has been expanding. There are several main sources used for learning analytics such as virtual learning environment (VLE), student information system (SIS), learning management system (LMS), and library system.

7.3.1.1 VIRTUAL LEARNING ENVIRONMENT (VLE)

VLEs are currently well known as part of education systems as ways to structure, manage, and provide learning exercises and the content. It is widely recognized for having the ability to track the student as well as manage online evaluation (Jisc, 2016). The VLE can be seen as an internet-based way of learning. This system will enable the student to access to learning materials, classes, and other resources online (Borwarnginn and Tate, 2014). The system can be accessed either on or off campus through internet technology and smart mobile devices (Susanto, 2018; Almunawar et al., 2015a; 2015b; 2018; Susanto et al., 2016). This implied that the system can encourage that student can learn outside the school as it can be accessible 24 h a day and even 7 days a week. This system is not only available for full-time students but also to those who cannot attend the campus regularly due to living at different geographic or time restrictions like those who are taking evening classes, distance learning courses, or part-time students.

The new learning process is more beneficial in comparison to traditional learning method. For instance, the interaction between teachers and students can be done through the internet. Student can just submit their assignment by uploading to the site and they can get instant feedback from their teacher. This can improve the communication between the students and the teachers. In addition, the course information including course outline, exam papers, course material, timetables, and other related information can be found in one designated website for this system. This can actually save cost and time as the teachers do not have to spend their money on printing the lectures notes, whereas the student can get the learning materials in advance. However, the limitation of learning through VLE tends to be less effective. This way of learning does not leave an impact toward the students due to the absence of face-to-face education. Even though virtual learning is said to be more interactive, the level of exchanging information is slightly different using computer (Arslanand Kaysi, 2013).

7.3.1.2 STUDENT INFORMATION SYSTEM (SIS)

SIS is an important part of education system especially in this modern era. It is believed that this system could solve the problem or difficulties using traditional approach of gathering relevant information of the student through the internet (Akiwate et al., 2016). Moreover, this system is design to replace the paper-based system of collecting and recording student information which is time-consuming and hard to manage (Bharamagoudar et al., 2013). SIS consists of students' detailed information in terms of their prior qualifications, socioeconomic status, ethnic group, modules choices, and grades acquired to date. From this system, there will be potentially valuable insights that could be used to designed VLE information so that it will be easier to predict student academic performance in education settings (Sclater et al., 2016).

The majority of the SISs isinvolved in administrative activities such as admissions, enrolment, and examinations. Apparently, the use of SISs can be beneficial for education institutions to record and maintain student information easily, coordinating different levels of student information, providing access to information by students, instructors, and parents 24/7, and storing biometric system information (Asogwa et al., 2015). Beside these, the system is user-friendly, provides good and efficient services, minimizes the workload of the users, and it is easy to retrieve data. The drawback of using the system is that it is hard to master the necessary coding as the control management of the system is using the latest technology, MEAN stack (Akiwate et al., 2016).

7.3.1.3 LEARNING MANAGEMENT SYSTEM (LMS)

LMS is widely recognized as the framework that manages every aspect of the learning procedure. The use of the framework of LMS supports in delivering and managing contents, identifies and evaluates an individual or organizational learning or training objectives, tracks the development toward those objectives, and gathers and displays the overall information of learning procedure. It also includes in managing course registration, administration, analysis, tracking, and reporting (Watson and Watson, 2012). The LMS is the most common data source for higher educational institutions through an online portal that connects with the lecturers and undergraduate students (Adzharuddin and Ling, 2013).

The learning process in higher educational institutions is no longer the same as in the primary and secondary school system which is only a one-way learning process. In higher education systems, the lecturers usually provide the lecture notes and extra information for the students to find out on their own. This means that in higher education system, students are encouraged to be exploratory learner throughout their studies. Moreover, throughout the lecture session, the lecturers will share their knowledge on certain topics and students need to participate by giving their own opinions or thoughts on the related topic during class discussion. Hence, university students need to continuously expand their knowledge by gathering more information.

The advantage of using LMS for lecturers is the ease of manage a large number of students in the system. This will also save their time as the lecturers can easily identify and monitor their students. For example, the lecturer has three different lecture classes. With LMS, the lecturer can easily know which students are taking his or her classes based on the classes the students are registered. In this case, it will be easier for the lecturer to give lecture material as well as receive assignments from the students. The drawback of using LMSisthe difficulty to have active discussion with all the students. For example, when there are hundreds of students posting messages to a lecturer, it is difficult for the lecturer to reply the student messages one by one. This can be time-consuming and increase stress level of the lecturer due to many messages.

7.3.1.4 LIBRARY SYSTEM

Transforming library into cloud-based shared system can actually contribute to the success of learning analytics. A vast amount of data could be generated through library system (Kiel, 2012). Basically, the library system has the ability to record student information, especially when the students are borrowing books from the library or when the students are accessing library system to read electronic journal. This piece of information is very useful for analytics to study the number of expected students visiting the library and student behavior.

There are a number of initiatives to enhance the library system in order to provide future generation learning tools. One of the initiatives is due to many educational institutions that are looking for distinct collection of sources for teaching purposes. This is where library system plays an important as the systems consist of diverse academic sources including thesis papers, past-year papers, presentation slides, articles, journals, reports, and other sources. Moreover, the source of information used for VLEs and LMS usually comes from the library system (Chad, 2015).

The advantage of library system is the ease of use. The students or lecturers can just search books through the system without the hassle of going to the library looking for books manually. Not only that, the system can be accessed 24 h and this makes it convenient for both students and teachers as they can access library sources any time they want. Moreover, digital copy of the books, journals, thesis, and other academic sources are available. Students and teachers can benefit from these as they do not have to carry heavy books from one place to another. The drawback of using library system is that when there is no internet connection, the system cannot be accessed.

7.3.1.5 MOST COMMON DATA SOURCES BEING USED IN THE INSTITUTIONS IN BRUNEI DARUSSALAM

In Brunei Darussalam, the most common data sources as a learning analytics that is being adopted by the institutions are the VLEs, SIS, LMSs, and library systems.

The VLE is being practiced by TelBru learning center, one of the telecommunication company in Brunei. By offering this kind approach of learning or training experience, it can be an advantaged way of learning things online in order to optimize the acquisition of know-how and capabilities. The design of the classroom at TelBru basically has three wide screens TVs in front of the room and a series of networked, high-definition screens on every individual work group member's tables. This kind of arrangement will aid the learners on how to work in a small group with the assistance of the mentor on the learning experience. Moreover, this new solution of cloud-based learning enables TelBru to touch the worldwide topics specialists live into the classroom (TelBru, 2015). International School Brunei also introduced this system in a way to enhance the student learning experience. The system is called the Student Advantage Program where the students were given free Microsoft 365 and storage space on cloud technology. This will allow the student to make full use of the Microsoft office to work together with other students and save their school work easily. By introducing this initiative, it will devise the students to have the right skills needed in the 21st century (Hajar, 2014).

A majority of the education institutions in Brunei Darussalam used SIS. The SIS for both primary and secondary schools is known as the Integrated National Education Information System (iNEIS). The reason for introducing the system is to take measure on implementing, monitoring, and measuring the current educational system, so that it can help the Ministry of Education to streamline and change certain business procedures in the effort to support the current needs for education. The system has detailed information that is needed for daily school operation which includes student recruitment, admission, attendance, assessment, examinations, school management, academic operation, transfer students, withdrawal, deferment, student allowances, teacher and student record, student discipline, clearance, school resources (scheduling and timetable), curriculum materials, student pass, invoices (billing), awards, cocurricular activities, scholarships and bursaries, and statistics and reporting (iNEIS, 2014). One of the aims of Wawasan Brunei 2035 or Brunei Vision 2035 is attempting to produce well-educated and highly skilled people by the year 2035; therefore, it is important to monitor the student performance. iNEIS system as the key performance indicator in order to monitor the performance of the student in terms of their attendance in school, school leaders and parents play an important in making sure in monitoring and supporting effectively on the student learning development (Thien, 2015).

In higher education institutions, the national university of Brunei Darussalam is implementing the use of LMS. Based on the Universiti Brunei Darussalam associate professor and the assistant vice-chancellor of academic affair, a LMS known as CANVAS[1] have been developed for both lecturer and students of Universiti Brunei Darussalam as it can be used to create online courses. The system not only assists in online learning but also creates opportunities for the working group to take up the courses due to its flexibility. This means that it can be seen that technology has the capability to contribute to teaching and learning in the near future (Yap, 2015). In terms of library system, most of the higher educational institutions in Brunei have their own library system so that it will be easier for the students to make full use of the library resources and provide the students with greater learning experience (Almunawar et al., 2013a; 2013b).

7.3.2 HOW DOES BIG DATA CONTRIBUTE TO LEARNING ANALYTICS?

The learning analytics has steadily caught the attention of education policymakers, leaders, and professionals. With the use of big data that can be utilized to customize each of experience users on business sites, education systems, organizations, and publisher's views, there is a huge potential in using the same data mining approaches to enhance the learning outcomes. The notion of using the data is by adapting instruction that meets with the individual learner need in real time similarly as Amazon, Netflix, and Google use metrics to tailor suggestions to customers (NMC Horizon Report, 2014). In this way, the potential of applying learning analytics can assist on changing certain areas of the education system from a standard one size fits all conveying system into a more interactive and flexible system that is created to meet the student academic needs and interests.

---

[1]A Learning Management System in University of Brunei.

These ideas have been a focal point of adaptive programming in which the programs can make calculation adjustments in order to keep the learners to stay motivated, especially the moment they are trying to master the concepts or experiencing difficulties.

7.3.3 CHALLENGES OF LEARNING ANALYTICS

Learning analytics is said to be offering greater opportunity in providing the best learning experience for students, especially in this big data era. However, there are few challenges that need to be taken into account, especially the ethical issues that are likely to occur.

One of the ethical issues of learning analytics that can be raised is the mislabeling of students based on incomplete or incorrect information or incorrect calculation. This factor need to be consider because the ways learners behave either in classrooms or through the internet depend on the level of complexity of personal, emotional, social, and economic factors that cannot be directly observed from their behavior alone. In this case, learning analytics may have lessened the effectiveness or limit to a learner's choice of access to specific materials or resources (Sharples et al., 2014). For instance, the adoption of learning analytics believes to suggest courses and modules with improvement in a particular student's chance of completing their qualification successfully. There will be students more likely taking easy courses with the highest chances of success than those students who are taking high level of difficulty and more challenging courses. In this case, it is essential to include students as active agents and collaborators. Through student-oriented learning analytics and the data being shown to the learners, it will give both lecturers and students the chances for self-reflection and the development of shared understanding.

Another ethical issues of learning analytics involve the legal issue specifically privacy concern, security, and personal information of the students and the teachers as this information are needed as part of educational data for learning analytics (Susanto and Almunawar, 2018). Both teachers and students may raise a question regarding the security of their personal information and who likely to access their information basically about their ability and knowledge. This issues need to be addressed as the students and the teachers definitely want answers on to what extent are their personal information is going to be used because it can be considered unethical if the information is misused and this can actually lead to violating the student and the teacher's personal information (Susanto and Almunawar, 2016; Susanto et al., 2011).

Apart from that, it should be highlighted that how exactly learning analytics can enhance the learning process in educational settings. In this case, it is important to note down certain criteria of learning analytics in terms of its reliability, validity, effectiveness, and usefulness in generating learning outcomes. The challenge in this area is basically what are the long-term effects of incorporating learning analytics in education setting contributing to learning processes.

Moreover, the challenge is that how much prepare is the education institutions inthe adopting learning analytics. This is because not everyone knows how to use the big data and learning analytics properly. Without proper guidelines on how to effectively use the information, it might affect the performance of the student adversely. It is crucial to make preparation by having expertise on this area and changing the infrastructure in order to ensure effective use of learning analytics in learning processes.

7.4 CONCLUSION

In conclusion, learning analytics is seen to have greater potential in delivering new process of learning in a way it can help improve the student performance in education settings. However, there are several challenges that might hinder the application of learning analytics. Therefore, it is important to address these issues so that learning analytics can be used effectively that suited with the 21st century of educational system.

7.5 RECOMMENDATION

As for the recommendation, there are certain areas that need to be improved in order to have better outcome of adopting learning analytics in educational institutions.

7.5.1 TALK ON BIG DATA AND LEARNING ANALYTICS

The first recommendation is by giving briefing to both teachers and students. Both teachers and students should be informed about what is big data and what learning analytics is all about because they are part of the learning process. By doing so, this can raise awareness about the benefit of implementing learning analytics in education setting and as a result, it can contribute to effective use of data for improving the student learning experience.

7.5.2 DEVISING ETHICAL FRAMEWORK FOR LEARNING ANALYTICS

The second recommendation is to devise appropriate ethical framework for learning analytics, especially education institution. It is important to define appropriate ethical measures for learning analytics so that the users will not misuse the information to the extent of violating other people personal information. In this case, it is very important to take appropriate measure when things can go wrong and give informed consent to the student so that they will understand that their information will be used only for educational purposes.

7.5.3 CONDUCTING A WORKSHOP

The third recommendation is by conducting a workshop. It is important to note down that not all people are computer literate because different people react differently to the level of complexity on the area of computer. For instance, teacher who does not have any basic of computer skills will find it hard to use it. In this case, it is recommended to give training step by step so that both teachers and students will be familiar on how to use it effectively and this can enhance learning processes.

KEYWORDS

- **educational institution**
- **learning process**
- **library system**
- **personal information**
- **virtual learning environment**

REFERENCES

Adzharuddin, N. A.; Ling, L. H. Learning Management System (LMS) Among University Students: Does It Work? *Int. J. E-Educ. E-Bus. E-Manag. E-Learn.* **2013,** *3*(3), 248.

Almunawar, M. N.; Anshari, M.; Susanto, H. Crafting Strategies for Sustainability: How Travel Agents Should React in Facing a Disintermediation. *Operational Res.* **2013a,** *13*(3), 317–342.

Almunawar, M. N.; Susanto, H.; Anshari, M. A Cultural Transferability on IT Business Application: iReservation System. *J. Hospitality Tourism Technol.* **2013b,** *4*(2), 155–176.

Almunawar, M. N.; Anshari, M.; Susanto, H.; Chen, C. K. Revealing Customer Behavior on Smartphones. *Int. J. Asian Business Information Management* **2015a,** *6*(2), 33–49.

Almunawar, M. N.; Susanto, H.; Anshari, M. The Impact of Open Source Software on Smartphones Industry. In *Encyclopedia of Information Science and Technology,* 3rd ed.; Khosrow-Pour, M., Ed.; IGI Global, 2015b, pp 5767–5776.

Almunawar, M. N.; Anshari, M.; Susanto, H. Adopting Open Source Software in Smartphone Manufacturers' Open Innovation Strategy. In *Encyclopedia of Information Science and Technology, 4th ed.;* Khosrow-Pour, M., Ed.; IGI Global, 2018, pp 7369–7381.

Asogwa, D. U.; Abdullahi, M.; Bello, A. The Need for Students Information Management System (SIMS) for Nigerian Universities in a Technological Age: Challenges and Strategies for Proper Integration, 2015.

Akiwate, B.; Patel, A.; Nabiwale, T.; Naik, N.; Patil, S. Web Based System Information Management System Using MEAN Stack, 2016. https://ijarcsse.com/docs/papers/Volume_6/5_May2016/V6I5–0226.pdf (accessed Sep 14, 2016).

Arslan, F.; Kaysi, F. Virtual Learning Environments. *J. Teach. Educ.* **2013**, *2*(4), 57–65. http://www.universitypublications.net/jte/0204/pdf/HVD867.pdf (accessed Sep 11, 2016).

Attwell, G.; Kieslinger, B.; Blunk, O.; Schmidt, A.; Schaefer, T.; Jelonek, M.; Reynard, C. Workplace Learning Analytics for Facilitation in European Public Employment Services, 2016.

Bharamagoudar, S. R.; Geeta, R. B.; Totad, S. G. Web Based Student Information Management System. *Int. J. Adv. Res. Comput. Commun. Eng.* **2013,** *2*(6), 2342–2348.

Borwarnginn, P.; Tate. A. An Investigation into Students' Perceptions and Lecturers' Perceptions of a Virtual Learning Environment. EDULEARN14 Proceedings, 2014; pp 1548–1554.

Chad, K. Library Management System to Library Services Platform. Resource Management for Libraries: A New Perspective, 2015. http://helibtech.com/file/view/Resource_management_briefing_HELibTech_KenChad_Aug2015.pdf (accessed Sep 16, 2016).

Hall, M. World Insight: The Global Potential of Learning Analytics. Times Higher Education, March 11, 2016. https://www.timeshighereducation.com/blog/world-insight-global-potential-learning-analytics (accessed Sep 13, 2016).

iNEIS. About iNEIS, October 22, 2014. http://ineis.moe.gov.bn/mod/page/view.php?id=1 (accessed Sep 12, 2016).

Jisc. Virtual learning environments (VLE), January 20, 2016.https://www.jisc.ac.uk/guides/technology-and-tools-for-online-learning/virtual-learning-environments (accessed Sep 12, 2016).

Kiel, R. Learning Analytics. Caul, 2012. http://www.caul.edu.au/caul-programs/teaching-learning/highered-resources/learning-analytics (accessed Sep 11, 2016).

McNulty, E. Understanding Big Data: The Seven V's. Dataconomy, May 22, 2014. http://dataconomy.com/seven-vs-big-data/ (accessed Sep 11, 2016).

National Science Foundation.Core Techniques and Technologies for Advancing Big Data Science and Engineering, 2012.http://nsf.gov/pubs/2012/nsf12499/nsf12499.htm (accessed Sep 11, 2016).

NMC Horizon Report. Higher Education Edition, 2013. http://www.nmc.org/pdf/2013-horizon-report-HE.pdf (accessed Sep 15, 2016).

NMC Horizon Report. Higher Education Edition, 2014. https://www.nmc.org/pdf/2014-nmc-horizon-report-he-EN.pdf (accessed Sep 13, 2016).

Picciano, A. G. The Evolution of Big Data and Learning Analytics in American Higher Education. *J. Asynchronous Learn. Netw.* **2012,** *16*(3), 9–20.

Rao, R.; Pravana, M.; Mounika, A. Effect of Big Data Characteristics on Security Leveraging Existing Security Mechanisms for Protection. *J. Eng. Appl. Sci.* **2015,** *10*(5). http://www.arpnjournals.com/jeas/research_papers/rp_2015/jeas_0315_1708.pdf (accessed Sep 12, 2016).

Scapin, R. Learning Analytics in Education: Using Student's Big Data to Improve Teaching, 2015. http://www.reptic.qc.ca/wp-content/uploads/2014/10/2015–04_Learning_Analytics.pdf (accessedSep 15, 2016).

Sclater, N.; Peasgood, A.; Mullan, J. Learning Analytics in Higher Education: A Review of UK and the International Practice, 2016. https://www.jisc.ac.uk/sites/default/files/learning-analytics-in-he-v2_0.pdf (accessed Sep 12, 2016).

Sharples, M.; Adams, A.; Ferguson, R.; Gaved, M.; McAndrew, P.; Rienties, B.; Weller, M.; Whitelock, D. Innovating Pedagogy, 2014. http://www.open.ac.uk/iet/main/sites/www.open.ac.uk.iet.main/files/files/ecms/web-content/Innovating_Pedagogy_2014.pdf (accessed Sep 12, 2016).

Hajar, S. ISB Promotes Virtual Learning Environment. Borneo Bulletin, September 27, 2014. http://borneobulletin.com.bn/isb-promotes-virtual-learning-environment/ (accessed Sep 10, 2016).

Sonka, S. Big Data Characteristics. International Food and Agribusiness Management Review, 2016.http://www.ifama.org/resources/Documents/Volume%2019%20Issue%20A/01_Editor.pdf (accessed Sep 13, 2016).

Susanto, H. Managing the Role of IT and IS for Suppoting Business Process Reengineering. *J. Systems Information Technol.* **2016a.**

Susanto, H. IT Emerging Technology to Support Organizational Reengineering. **2016b.** https://ssrn.com/abstract=2770318

Susanto, H. Cheminformatics—The Promising Future: Managing Change Of Approach Through ICT Emerging Technology. In *Applied Chemistry and Chemical Engineering, Volume 2: Principles, Methodology, and Evaluation Methods*; Haghi, A. K., Pogliani, L., Balkose, D., Mukbaniani, O. V., Mercader, A. G., Eds.; Apple Academic Press, 2017a, pp 313–332.

Susanto, H. Biochemistry Apps as Enabler of Compound and DNA Computational: Next-Generation Computing Technology. In *Applied Chemistry and Chemical Engineering, Volume 4: Experimental Techniques and Methodical Developments;* Haghi, A. K., Pogliani, L., Castro, E. A., Balköse, D., Mukbaniani, O. V., Chia C. H., Eds.;2017b, Apple Academic Press, pp 181–202.

Susanto, H. Electronic Health System: Sensors Emerging and Intelligent Technology Approach. In *Smart Sensors Networks;* Xhafa, F., Leu, F.-Y., Hung, L.-L., Eds., Academic Press, 2017c, pp. 189–203.
Susanto, H. Smart Mobile Device Emerging TECHNOLOGIES: an Enabler to HEALTH Monitoring System. In *High-Performance Materials and Engineered Chemistry;* Torrens, F., Balköse, D., Thomas, S., Eds.; Apple Academic Press, 2018, pp 241–264.
Susanto, H.; Almunawar, M. N. Security and Privacy Issues in Cloud-Based E-Government. In *Cloud Computing Technologies for Connected Government;* Mahmood, Z., Ed.; IGI Global, 2016, pp 292–321.
Susanto, H.; Almunawar, M. N. *Information Security Management Systems: A Novel Framework and Software as a Tool for Compliance with Information Security Standard.* CRC Press, 2018.
Susanto, H.; Chen, C. K. Information and Communication Emerging Technology: Making Sense of Healthcare Innovation. In *Internet of Things and Big Data Technologies for Next Generation Healthcare;* Bhatt, C., Dey, N., Ashour, A. S., Eds.; Cham: Springer, 2017, pp 229–250.
Susanto, H.; Chen, C. K. Macromolecules Visualization through Bioinformatics: an Emerging Tool of Informatics. *Appl. Phys. Chem. Multidiscip. Approaches* **2018a,** 383.
Susanto, H.; Chen, C. K. Informatics Approach and Its Impact for Bioscience: Making Sense of Innovation. *Appl. Phys. Chem. Multidiscip. Approaches* **2018b,** 407.
Susanto, H.; Almunawar, M. N.; Tuan, Y. C. Information Security Management System Standards: a Comparative Study of the Big Five. *Int. J. Electrical Computer Sci.* 2011, *11*(5), 23–29.
Susanto, H.; Almunawar, M. N.; Leu, F. Y.; Chen, C. K. Android vsiOS or Others? SMD-OS Security Issues: Generation Y Perception. *Int. J. Technol. Diffusion* **2016,** *7*(2), 1–18.
Susanto, H.; Chen, C. K.; Almunawar, M. N. Revealing Big Data Emerging Technology as Enabler of LMS Technologies Transferability. In *Internet of Things and Big Data Analytics Toward Next-Generation Intelligence;* Dey, N., Hassanien, A. E., Bhatt, C., Ashour, A. S., Satapathy, S. C., Eds.; Cham: Springer, 2018, pp 123–145.
TelBru. TelBruLearning Centre: the First of Its Kind in Brunei, October 27, 2015. https://www.telbru.com.bn/page.php?c=309 (accessed Sep 12, 2016).
Thien, R. MOE: We'll Monitor School Attendance in Real Time. Brunei Times, February 10, 2015. http://m.bt.com.bn/news-national/2015/02/10/moe-we%E2%80%99ll-monitor-school-attendance-real-time (accessed Sep 11, 2016).
Watson, W.R.; Watson, S. L. An Argument for Clarity: What Are Learning Management Systems, What Are They Not, and What Should They Become? 2012. https://halshs.archives-ouvertes.fr/hal-00692067/document (accessed Sep 14, 2016).
Yap, A. Online Learning in the Pipelines. Brunei Times, February 15, 2015. http://www.bt.com.bn/news-national/2015/02/15/%E2%80%98online-learning-pipelines%E2%80%99 (accessed Sep 11, 2016).
Zikopoulos, P. C.; Eaton, C.; Deutsch, T.; Lapis, G. *Understanding Big Data: Analytics for Enterprise Class Hadoop and Streaming Data.* McGraw Hill:United States, 2015.

CHAPTER 8

BUSINESS INTELLIGENCE AND ANALYTICS: A BIG DATA PARADIGM

HERU SUSANTO

The Indonesian Institute of Sciences, Jakarta, Indonesia

Tunghai University, Taichung, Taiwan,
E-mail: heru.susanto@lipi.go.id

ABSTRACT

Currently, the most interesting subject discussed within the business community internationally is "big data." Thus, the era of big data is born where a huge amount of raw data or information can be obtained and processed. This has enabled business managers to open new doors that could lead to the accomplishment of high pertinence and effective technological changes (paradigms). The rise of new informative accumulation technologies through certain analytical tools and softwares, has propelled information mining and investigation support and guides exploration strategies for management to use to their advantage. This surely aligns with the main objectives of the big data, where it helps management to have a diverse business view, which reflects data of the past and current records that analyze the organization performance.

This report will focus on how the big data paradigm has influenced the way businesses utilize business intelligence (BI) and analytics, how top managers have the capabilities of accessing vital information, and how to process the analysis quickly using sophisticated tools and software. It will also discuss what are the other issues that the paradigm shift of big data has brought, such as ethical matters on whether or not the action of these analysts in obtaining data with or without the knowledge of the targeted customer is considered right or wrong. This report will further analyze who would benefit from these shifts and does it give any leverage for any business in utilizing this huge amount of data against their rival competitors. Moreover, this report will analyze what other issues may arise due to this technological advancement of big data such as ethical issues.

8.1 INTRODUCTION

The rapid evolution in modern technological development, especially in the field of information technology (IT), it has been definitely defined the way, we conduct our daily activity. These technologies have a profound impact toward the business world as IT today can be obviously seen influencing the way large business organization makes business decision. This evolution of technology that has created larger information data capacity have somewhat burst into the scene (the business world in general) in the 21st century, with those organization who were the first to adapt these new technologies have managed to gain some leverage against their competitors.

Some of the most well-known organizations such as Google and Facebook were born and developed huge information from the earliest starting point. They did not need to combine enormous information innovations with their conventional IT frameworks on the grounds those bases did not exist. Big information could remain solitary and enormous information investigation or analysis, could be the main center of analyzing. A largeamount of information innovation structures could be the main engineering driving force that could pave the way for business organizations to use as one of their strategic tools to increase their chances of success.

8.2 LITERATURE REVIEW

8.2.1 PARADIGM SHIFT OF BIG DATA

As a result of a dynamic evolution of today's modern technology which has become more sophisticated and complex, it is now possible for people to do amazing things such as in this context, amassing huge quantity of data information which isanalyzed and interpreted rapidly through such technology. And it does not limit to that, with this complex technology, people can now find certain behavioral pattern based on data that enables to make accurate prediction. These changes or shifts in technology or in the common term that will use this report as a big data paradigm, have an astounding impact toward the process or method on how big organization (in this case will refer mostly to the business world) handles their business intelligence (BI) and analytics activities.

These changes in the technology have enabled the companies or organizations the ability to manipulate and process a huge amount of raw data as quickly and as efficiently as possible. This is made possible through the development of certain application of software's and tools. The businessorganization used to store and retrieve data through data warehousing which may take longer time and involve more cost to the organization. With the advancement in technology especially in IT, it has given top management the ability to access data/information at the tip of their fingers and a speed rate that could prove to be vital for any critical business decisions that need to be made quickly and more importantly as accurate as possible.

Throughout the time, as these data stored in data warehouses keep on accumulating in mass sizes, it has become extremely difficult for those who are looking for a particular data which resulted in more time-consuming process in order to search and locate. In business, time is considered to be so precious and business can lose huge amount of money due to the activity of searching required data which takes too long. By the time the company found the needed data, their competitors might have already seized the opportunity first.

In a worst-case scenario, business organization's top management may make wrong decision based on the wrong data that is being obtained. Because there is a huge amount of data to search from, those responsible to analyze these data get confused and erroneously retrieve and analyze the wrong data.

As big data technology has evolved, and with recent technology, the term data mining comes to play in which it can help to solve the problem of analyzing huge amount of data. Data mining is the activity in which analyzing and breaking down of data from alternative point of view and perspective and come up with some summarize data that is meaningful to the user. With this information obtained from data mining, it can provide the organization with critical information that can be used to increase profits, cut unnecessary cost, and even to some extent do both.

With the astounding results brought in by data mining, it has become one of the most popular methods or tools that big business organizations use to make critical business decision. According to Fayyad et al. (1996), recently there is a need for a new conceptual method of computational theories, advanced tools to guide and assist people in taking out useful data from the ever rapidly developing volumes of digital information.

8.2.2 BIG DATA

Recently, the term "big data" has been one of the best topics among top management within the business community and also for those who are in the scientific research industries. As a matter of fact, based on the research conducted by Yakabuski (2013), pointed out that IBM generates roughly about 2.5 quintillion bytes of raw data every single day. The potential of this huge amount of data toward the business world, are very precious as if it is processed correctly, this could ensure their business survival and may even give them some leverage against their competitors.

Unfortunately, most people still have a vague understanding of what it really means or what it really is. The reason behind this is probably because the term big data has a variety of definition from different people. In other words, different people have different perception on what the term big data means. This is why when one is asked to explain the term, they will have a difficult time to point out what it really means. When askto a group of people, let say about 10 people, you would expect to get 10 different answers from them on the definition of the term big data.

One of the most well-accepted definitions for the term big data is stated by Madden (2012), defines the term as having a huge amount of data (organization increasingly deal with petabyte-scale data sizes) which are obtain extremely fast which makes it quite impossible for traditional tools to process it.

Another view or definition regarding the term is made by Apache Hadoop, where their version define big data as datasets which is impossible for normal computers to process it within a time frame that is acceptable (Chen et al., 2014).

What can be understood from the above given definition of "big data," it can be said that there is no clear one definition of the term, but it is in fact as general term or it acts as symbolism like an "umbrella" that consist of many processes, procedures are blend together with certain sophisticated software's to help and guide top management to make critical business decision through complex analytic calculation to be able to predict future outcomes based on analyzing past records. Big data simply means the huge collection of raw data that requires modern state of the art technologies that have the necessary tools and softwares to be able to process such huge amount of data efficiently.

8.2.3 BUSINESS INTELLIGENCE (BI) AND ANALYTICS

BI and analytics have recently become one of the key crucial elements especially in the business world as it can definitely help top management to make critical business decision both rapidly and as accurately as possible. Most big companies nowadaysutilize some sort of BI and certain analytic tools to help them make a quick business decision, whereas those who do not take advantage of BI and analytics will definitely face a harder time competing with their competitors.

In other words, it refers to the strategy of the organization through utilizing IT software's to identify, extract, and analyze the raw data into some meaningful information that are related to business such as sales transaction records. BI can generally answer some important question such as what had happened, how many did happen, how frequent it happens and analyze where exactly the problem is and what type of actions are required to deal with it.

On the other hand, analytics means or is referred to as having a set of skills, utilizing sophisticated technology applications, and the practices for relentless investigation of past business performance to gain insight and encourage the development of business planning. Some of the data that is analyzed here will help to answering questions such as why is this happening, what will occur if these trends are made to continue and what are the company's optimal outcome in certain situations. The most attractive features or ability that today's big data technology can provide to business organization is that big data can help top management look into the future through analyzing past and present data and make some predictive analysis of what will happen and could be happening in the future.

Even though BI is among one of the best subjects currently, surprisingly, business intelligence is not actually a new thing, but in fact have been around a very long time but possibly with different terms used than that of today. This is supported by a research conducted by Chen et al. (2012), where they stated that BI and analytics have been continuously becoming more important for business community for the past two decades. The word "continuously" gives the picture that BI is still an ongoing process or work in progress that keeps on improving and developing through time.

Therefore, existing for so long, it has definitely evolved through time, in the beginning traditional method of gathering BI and analytics might be from obtaining data manually such as making the customers or people to give information and some IT expert would key it in and storing those huge data in a place called data warehouse. Not only this takes more time to search for the data when required, but also it is prone to human error that limits the amount of raw data, as the IT infrastructure are not capable to store and process a huge amount of data used like today.

8.2.4 "BIG DATA" A CURRENT TREND

In recent times, it has become a common trend for almost every business organization to use some kind of BIsoftwares/tools to gather vital information about their target market, toanalyze these data which will guide/help top management to make quick and accurate business decisions.

IT has become so advanced today that it has created a new trend among the community generally and the business world specifically. According to Susanto (2018) and Cukier (2013), they stated that "big data" is a great revolution of the most fascinating trend in innovation and the significant effect it will have on the economy, science, and society. This trend is so great in terms of giving the ability to crunch a vast quantity of raw data and being able to analyze and provide amazing results. The ability to obtain-huge amount of information will change the way business is conducted, well-being, governmental issues, training, and development in the years to come.

As a result of this trend, organizations are now looking for more managers and employees who have the necessary skills and knowledge to be able to retrieve huge amount of data to analyze it effectively. Unfortunately, most organizations tend to find it difficult to find such workers as there are a lot of organizations looking for them. Based on a study conducted by Lohr (2012), statistics showed in a report made by McKinsey Global Institute, it was anticipated that in the United States alone, will require a large number of employees who possess profound analytical skills with more than 1.5 million more data-literate managers needed.

From the above, the changes of big data technology have made BI and analytics quite popular nowadays and most business organization are high depended on this technology as it can help their business to be more successful.

8.2.5 ETHICAL ISSUES BROUGHT BY BIG DATA PARADIGM

Having so much power in one's hand comes with a very high sense of responsibility to utilize these powers in a more ethical manner. In relation, to what the big data paradigm to this statement is that having the capacity or ability to analyze such a huge amount of data may introduce issues concerning with ethics. For example, one should ask the question of what are the moral contemplations of business organization in utilizing huge information to target individuals without their insight or knowledge for the purpose of obtaining personal information for profits? Are these action made by the organization considered to be morally just to do such action and are the result far outweigh the moral principle in order for the organization to get what information are needed.

Indeed the concern for ethics is one of the most crucial factors and issue that every organization must consider in order to avoid any unnecessary negative implication that big data may bring toward the organization. Mateosian (2013), in his view regarding the topic of ethics in big data, pointed out that for a long time society has battled with the ramifications of gathering and analyzing personal data. The limitlessly expanded velocity and quantity of such activities have made a subjectively new circumstance (ethical issues), for the most part known as big data.

A significant portion of the hardest inquiries organizations are confronted today emerge out of this enormous information or better recognize as big data. For example, with enough data about the customer's surroundings or environment, the potential of somebody being able to know a great deal about them without knowing their name is a reality. However, by joining or linking such troves of data from divergent sources, an organization can construct an intrusive dossier about the customer without really abusing the well-intentioned privacy policies of the associations that initially gathered the information.

From what can be deduced from the above information, the evolution of big data hascreated some ethical situation that could have a profound impact toward any business organization, if they do not handle it with care and consideration. As consumer nowadays has become more knowledgeable and more conscious of what they spend their money on or in particular how businesses obtain their personal information, they would become more critical in the activities of any business organization and would start to question the activity ethically. The activity of obtaining personal data would certainly be one of the most sensitive issues that customers will take notice and would make a great deal about it which will force organization to react to this situation strategically and cautiously and to balance satisfaction between their stakeholders and customers.

8.3 DISCUSSION

The following discussion will initially discuss about the characteristics of big data and how the changes in technology affected it. It will then continue to discuss how the paradigm or in other words the shift of technology in regards to big data technology have impacted specifically toward the business world. It will look at different perspective of what these changes have brought about both positively and negatively as well as other issues that may arise because of these changes in technology.

8.3.1 MAIN CHARACTERISTIC OF BIG DATA

Before one can completely understand what big data really is and what are the possibilities that big data can give the organization, it is critical to know what the key characteristics of big data are. The following points will elaborate the main characteristics of big data that is in relation to velocity, volume, variety, veracity, and finally value.

8.3.1.1 VELOCITY

Velocity is the pace of data creation which is even more important than volume. The real-time or actual time data make it achievable for an organization to be faster than its competitors. According to Edosio (2014), velocity means how fast the data is produced which can be caught either constant, in batch, or in timeframe. For example, web analytic sensor normally captures the number of clicks on the website, by using particular programming functions which listening to the click event on real-time basis. For each click, the web sensor analytic is quickly updated. However in some cases, where the data was captured in batch such as bank daily transaction data will be review at the end of each day. Another example, the New York stock trade gathers 1 TB of information in a single trading session and having present information and real-time rules for trades and predictive modeling is essential for managing stock portfolios.

8.3.1.2 VOLUME

Volume is the amount of data created and stored. The size of the data decides the worth and likely understanding if it can be viewed as a big data or not. According to Chang (2016) as data agent, currently every day 2.5 quintillion bytes of data were produced, and extraordinarily, 90% of every single current data that is evaluated to be at 2 ZB was created just in the most recent 2 years with the vast majority of this data being unstructured. Different key occasions and trends have contributed enormously to the continuously increase in the volume of data which some of the trend consist of social networking, growth of transactional database, and increase in multimedia content (Edosio, 2014).

First social media which there has been significant development in the amount of data created from social networking site. For example, a normal people that use Facebook makes more than 90 content in a month. Additionally, every day there are around 35 million updating their status on Facebook. This is only a little picture considering that there are several applications that can use for social media which encourage users to interact with each other and share content day by day such as Instagram, WhatsApp, Twitter, and many more. Second, the growth of transactional databases where businesses are forcefully catching customer-related data, in order to analyze the behavior of the customer so that the business can improve their marketing strategy. Third, increase in Multimedia Content, currently media information represents more than half of all Internet traffic. Wielki (2013) stated that as indicated by the Internet Data Enterprise, the quantity of mixed media content developed by 70% in the year 2013.

8.3.1.3 VARIETY

Variety is different kind of data which both structured and unstructured data in many forms and combination need to be considered. According to Hashem et al. (2015) variety refers to the various sort of data gathered through sensor, signal from Smartphones, or social networks. Such data types consist of video, picture, text, audio, and data logs that can be in structured or unstructured format. For example, web journal, instant message, online games, and online networking produce different type of unstructured data through Smartphone device and sensors. Web users also produce an extremely diverse set structure data and unstructured data.

8.3.1.4 VERACITY

Veracity which represents the instability and unreliability inherent in some origin of data. For example, customer sentiments in social networking are indeterminate in nature, since they involve human judgment. Yet they contain valuable data. Thus, the need to manage with imprecise and indeterminate data is another aspect of big data, which is tended to using tools and analytics developed for administration and mining of uncertain information (Gandomi and Haider, 2015).

8.3.1.5 VALUE

Oracle presented Value as a characterizing trait of big data. Based on Oracle's definition, big data are regularly portrayed by relatively "low value volume." That is, data got in the first shape, for the most part, has a low value in relative to its volume. However, a high value can be acquired by analyzing vast volumes of such data (Gandomi and Haider, 2015).

8.3.2 CHALLENGES BROUGHT BY BIG DATA PARADIGM

Evolutionof big data has brought up a few challenges that organization has to face. Big data are important for them and if they cannot face up to these challenges there is a high possibility for them to not able to fully optimize the benefits of big data and to some extreme point, it fails completely.

8.3.2.1 ETHICAL CONSEQUENCES BROUGHT BY THIS BI AND ANALYTICS PARADIGM

Ethics have always been one of the key concerns of society especially toward activity conducted by large business organization. Consumers are always skeptical toward business organization that requests personal data from them. This issue is then heightened due to the development of BI ability to obtain the customer's personal detail accurately and in a discrete manner.

They argue that when business organization retrieves their personal detail, they can understand their (customer's) buying behavior or pattern. With this information, business organization can somehow able to manipulate customer into buying their product. This action to customers, clearly violates the moral ethics to them these organizations have no rights to obtain personal data and worse use that data to seek greater profits through data manipulation.

A research done by Smith et al. (2013) quoted that a large portion of customers are being impacted (influence) and controlled by manipulation, very significantly more than they realized especially in their behavioral pattern of the customer's consumption. Moreover, most of these organizations made extensive efforts with mostly impressive results, to change the customer's way of cognitive behavior which also influence the buying decision process, through the extensive retrieval, processing, and manipulation of raw data.

From the customer's point of view regarding ethical behavior of business organization taking personal information such as browsing history, they may view this action as morally or ethically wrong. As to them (customer), no consent or permission was given to business organization telling that they (business) can take these data without acknowledging them (customer) first of all.

On the other hand, from the perspective of the business organization, they may view moral issue differently. To them, it is not a problem or doing something wrong, so as long as it bring benefits to their customer as a whole. The business organization may use the Utilitarian theory or moral behavior to justify their action of getting these personal data from their customers. Utilitarian theory is judgments based on consequences. The main purpose or objective of the utilitarian theory is the result of the action should bring the greatest satisfaction as a whole.

According to Boddy et al. (2008), utilitarianism is someone who follows this methodology would consider the effect of an action of overall or general human being and would see an act as ethical if it creates more pleasure than pain. If it hurts a few people but benefits all then more, then it is moral.

Hence, based on the theory of utilitarian, from the business organization perspective, they might consider their action to be morally good as the activity of obtaining these personal data will give their customers benefits in the future. For example, through collecting and analyzing customer's data, the business organization can predict certain buying behavior or patterns. With this analysis, they can now produce better goods or services to be served to their customers. Here, it is clear how through obtaining personal data from customers can benefit a lot more customer in the future.

Therefore, as long as the activity of the business organization in collecting or taking personal data from the data does not harm them in terms of financially and they do not exploit the personal data, then according to the business organization, it is not an issue of conflicting moral behavior.

Negative consequence brought by these BI and analytic paradigm in terms of the company are hacking other company database in order to get the secret of rival company which is illegal and unethical. According to Hemmatfar et al. (2010), it is insufficient just to gather information from the competitor. Analyzing the data is as important as gathering it. They propose that one can utilize IT tool ranging from intelligent agent to data mining. Another, viler, part of competitive intelligence is industrial espionage where the company spying directed toward finding the secret of the rival company.

Other negative consequence can happen when there is rise of conflict of interest which the worker of the business uses the BI for their own personal gain. Recent cases happen at United States this year, where a worker at the Richmond Department of Social Service misuse their power and accessed computer to pull someone's personal information for their own benefit (Wise, 2016). Hashem et al. (2015) state that recently some controversies have uncovered whereby several agency use data generated by people for their own personal gain without any consent.

8.3.2.2 FINDING SUITABLE MANAGERS/WORKERS WITH THE REQUIRED SKILLS AND KNOWLEDGE

Unfortunately, these changes brought by technological advancement brought up several issues and challenges for organization to deal with an appropriate manner. One of the main difficulties that most businesses are facing is due to the emergence of big data, that is, the limited availability of staff or workers possessing a high IT and analytical skills and vast knowledge, in order to be able to retrieve and process these huge information/data and give a meaningful information for the organization in order to make a sound and accurate strategic decision.

This issue was mentioned in a study conducted by Chang et al. (2014), according to their research, there are extensive difficulties or extreme challenges for organization to overcome in order for them to really gain the full potential of utilizing big data completely. These challenges come in the form of managerial level that lacks the necessary skill and talent of a good analyst as it is hard/difficult for them to locate and identify the right people for the job. They continue to argue that in America alone, organizations are confronted with the lack of approximately 140,000–190,000 employees with profound analytical skills, whereas 1.5 million more managers who are analysts and are capable to make effective strategic decision.

In the perspective of labor force, the evolution of big datahas created a phenomenal situation especially in America, where organizations are struggling to find qualified managers and workers with specific skills to big data. Even if an organization has the technology to obtain the huge amount of data, without adequate skills and knowledge, they would not be able to retrieve important data and analyze it properly. The evolution of big data was significantly fast that business organization could not catch up in term of hiring suitable employees. However, most probably business organizations have only just realized how BI and analytics could help to improve their organization through providing vital information in order to make precise business decision and strategies.

8.3.2.3 USING REAL-TIME BIG DATA ANALYTICS

Real-time big data analytics is the utilization of, or the capacity to utilize all accessible enterprise information and resources when they are required. It is consist of element analysis and reporting, based on data went into the system less than a minute before the actual time of use. Barlow (2013) stated thatreal-time implies that the company processing the data in present, rather than process the data in the future. But "the present" also have other meanings to different users. From the point of view of online retailer, "the present" means the capacity to focus on a potential customer. If the processing time of transaction surpasses the customer attention span, the online retailer does not consider it as real time. From the point of view of optional trader that real time means milliseconds. In other words, real-time indicate the ability to process data or information as it arrives, rather than storing the data and retrieving the data later in the future (Barlow, 2013).

The example of tools that offer the possibility of real-time processing of big data is Apache Storm, Cloudera, Gridgain, and SpaceCurve. Apache storm which is currently owned by Twitteris a real-time distributing computation system. Apache storm uses the similar technique like Hadoop support a batch processing as it uses a set of general primitives for performing real-time analyzes. Apache storm is easy to utilize and it is compatible with any programming dialect. Cloudera deliver the modern platform for data administration and analytics. Cloudera also give the world's quickest, easiest, and most secure Apache Hadoop platform to help the customer solve the most difficult business issue with data. Cloudera also offers the highest performance on a low-cost platform for utilizing data to drive better business results. Powered by the world's most well-known Apache Hadoop distribution, Cloudera Enterprise gives the company a chance to concentrate on the result.

GridGain Systems, Inc. creates and gives in-memory programming. The GridGainin-memory data fabric is intended to control today's fast data challenges and unleashs the upper hand of any real-time business, either on-premise or in the cloud. GridGain offering the most comprehensive, venture grade in-memory computing solution for real-time analytics and hybrid data processing, high-volume transaction, GridGain allow fortune 500 organization and inventive mobile, web, and Software as a Service companies to expect and innovate ahead of market changes. Gridgain allows data-intensive application to scale out to a huge number of transactions per second while decreasing the query times 1000x or more without changing the database. Gridgain customers are in financial service, programming, e-commerce, medical service, online service, telecom, and more.

SpaceCurve is the technology that creates and finds underlying patterns in multidimensional geodata. Geodata is a diverse data than normal data as mobile device make new data very fast and not in a way old databases are used to. SpaceCurve offers a big data platform and has set a new world record in 2013 about running complex queries with tens of gigabytes per second.

Real-time big data analytic not only can providebenefit to business but also offers some challenges. First, real-time big data analytics requires a special computer or PC power. The current standard version of Hadoop has not reached its full potential in real-time big data analysis. At the moment Hadoop is not yet applicable for real-time analysis. New tools need to be purchased and utilized. There are however a few tool or device accessible to carry out the job. In the future, Hadoop will have the capacity to process data in real time. Second, using real-time insight requires an alternate method of working within the organization. If the organization regularly receives insight once a week, which is very normal in a lot of organizations, receiving this insight every second will require an alternate methodology and method of working. Insight requires action, activity, or movement and rather than acting on a weekly basis this action is currently in real-time required. This will have an effect on the way of life and on the culture. The objective should be to make the organization a data-driven organization.

8.4 EXAMPLE OF COMPANIES WHO ARE SUCCESSFUL TO EMBRACE BIG DATA EVOLUTION

There are a lot of companies and big business organization who have succeeded quite progressively in utilizing the technology that big data have given them. Through big data, these companies have managed to grow at an outstanding rate and success. All these companies are concrete evidence that proves through the evolution of big data, it can provide advantages to those organizations that have openly embraced it. Some example of big organization or business that is made successful through big data are Company One and Amazon.Com.

8.4.1 COMPANY ONE

Capital One was established by Mr. Richard D. Fairbank in 1988 based on his conviction that the power of data, innovation, testing, and awesome individuals could be joined to bring highly customized financial product specifically to the customer. From that point forward, Capital One has developed as one of the America's biggest consumer franchises with one of the country's most perceived brands. As one of the country's main 10 biggest banks based on deposit, Capital One, NA has branch areas fundamentally in New York, New Jersey, Texas, Louisiana, Maryland, Virginia, and the District of Columbia.

Capital One has been tackling behavioral information to shape customer offerings for a considerable length of time. For instance, their deal improvement engine analyzes customer demographics and spending examples to decide how, where, and when to place offers in front of the people. Which it lead to more income for Capital One and a more positive experience with the brand for the customer. They also made Capital One Labs, a kind of tech-driven research organization where employees utilize big data to deal with potential opportunities such as new methods of portable managing an account or mobile banking. Company one has successfully changed from a small bank division to a new status as world-class fortune 500 company.

8.4.2 AMAZON.COM

Amazon.com was founded by Jess Bezos in 1994 and in 1995 Amazon open it's website. Previously, Amazon was an online bookstore which only had 400 title books. But in 2000, Amazon started to grow as they raise their product to 4.7 million books, music, movie, and many categories. Having realized that the huge amount of information Amazon had about their customer, could be put to valuable use, they got to be one of the earliest adopters and using big data, went ahead to change the way they did business. Amazon uses the big data to maximizing the business benefit by patenting the shipping of goods, improving customer care quality, use the 360-degree customer profiles, preventing theft and Amazon Web store (AWS).

Amazon patenting the shipping of goods known as anticipatory shipping. Every time the customers browse through Amazon web page, the retailer will collect a little bit of data about their customer. After a year, Amazon will know well about their customer. With the data, Amazon uses the system for sending an item to customer when Amazon think that the customer might order it even when the customer not yet click the button. The anticipatory rely on the product that was purchased by the customer before and also other factor such as the wish list content and the previous browser by the customer. Moreover, the system helps in term of reducing the wait time for the product. The 360-degree customer profile will give Amazon the view and better insight into the behavior of customer as it happens. Because of that not only Amazon master the recommendation of their product such as book, toy, music, or kitchen utensils that their customer might want to order it or interested in but also they use the big data to improving customer care quality.

By using the big data, Amazon can prevent thief as Amazon have more than 1.5 billion items in its catalog and there are more than 200 fulfillment centers around the world. That is a lot of objects in a ton of places for the online retailer to monitor. Keeping the most valuable things secured is not as simple as putting the highest priced items under lock and key. Sometimes, because of limited accessibility or other factors, a lower value item might actually be more targeted by criminals. To figure out which product are most likely to be stolen, Amazon store the item inventory data in Amazon Simple Storage Service which end up having more than 50 million updates every week. Every 30 min, the team spins up Amazon compute clusters, crunch the data, and the data is fed back to the distribution center and site. Amazon also expands the business which maximizes the big data by offering AWS, where the customer can use the server, store the data, and use the application over the Internet.

8.5 BENEFIT AND DRAWBACK OF BI AND ANALYTICS A BIG DATA PARADIGM

The changes of technology will always bring in both positive and negative implications. The development of big data technology also brought in some benefits as well as some drawbacks toward the users. It is up to the organization to carefully weigh in both of these two factors and decide whether or not it is suitable for them to consider to use in their organization.

8.5.1 BENEFITS

First, business organization can gain several benefits through the development of big data not only financially in terms of gaining more profits but it can also make business operation to run more effectively and efficiently as possible.

Decision-making: One of the most significant advantages that big data can give is in terms of decision-making. In the business world, the ability to make accurate decision and as quick as possible may be one of the determining factors that contribute to a successful business. Unfortunately, decision made by human beings are constantly prone to mistake and sometimes these mistakes can be extremely costly for the business. These mistakes might have been made due to the limitation of period they have to make or decide on some strategies.

For humans, it is normal to take some time in deciding what strategies to use as it requires some deep analysis before one can come up with a decision. Even though sometimes some business organizations can get away with it; unfortunately, most will suffer greatly because of this as in business time and accurate decision is money.

With the presence of BI and analytics technology, it has somehow changed the game of how business organizations are able to make decision in the early phase of development, the most typical method of businesses to collect BI is through data warehousing. In the past, this method was quite manageable as during that early period, the amount of data collected was not that great.

Unfortunately, with the advancement of technology, more modern and sophisticated tools and software has made BI and analytics more efficient. This has greatly helped business to make quick decision. Especially, the current amount of data is now very huge and if business uses the old method it will take some time. These modern tools and softwares have made it possible for analyst to retrieve, analyze, and interpret huge amount of data at a very high speed. This would definitely be helpful for managers to make quick strategic business decisions. The faster they make decisions, the better for the organization to compete with their rivals.

Being able to make quick decision alone does not guarantee it will give a good decision. This is where the modern technology of BI and analytics come to play. With the current tools and software, managers will now be able to collect more precise and reliable data for the organization. BI and analytical tools currently have the ability to analyze past data, analyze the data, and give some suggestion on why things are happening and when is it happening. It can also suggest some ideas on how to deal with the issue and find some possible solution.

And most importantly BI and analytical softwares can predict future outcome. With this ability, not only does the manager access vital information and come up with quick decision but also the data given is as accurate as possible. Thus, BI and analytics software can lead to the organization in improving knowledge management, decision-making of the company, increasing the knowledge on marketing, and sales skill.

BI and analytical tools and softwares are not only beneficial for the business world, but also it benefits other sector such as health sector in which it reduces the amount of fraud waste and abuse. Through using the predictive analytics, it helps them to detect such fraud effectively and quickly. According to Tahir (2015), the content management system, one of the BI tool, it has either managed to identifiy or avoided more than $210.7 million in healthcare fraud in 1 year since utilizing the predictive analytics system.

In brief, some of the processes of predictive analytics are to allow risk scores to a particular claim and provider, to recognize billing pattern and claim aberrancies which were extremely difficult to distinguish by using the traditional methods. Bigdata also improves the outcome as the detail of patient information is increasing rapidly which will mean more detail and up to date information. Besides being able to store the patient information, big data combine and analyze a variety of structured and unstructured data from numerous data sources. With this amazing ability, it can increase the precision of diagnosing patient conditions, matching treatments with result, and foreseeing patients who are at risk for disease in the future.

Using Real-Time Big Data Analytics: Real-timebig data analytics the advantages are enormous, the error occurs within the business or organization are instantly aware of staff. Real-time insight into error or glitch helps the organization respond quickly to check the effect of an operational issue. This can recover the operation from falling behind or entirely fall or it can spare the customer from having to stop utilizing the company products. Recently, Samsung company recall back their new product Samsung note 7, so that it can minimize the damage and safety to Samsung company and customer. According to Azli (2016), Samsung acknowledges 36 cases of reported battery blasts globally which the recent one happened in Australia lead by the flaw in the battery. Therefore, Samsung company offers to repair, replace, or refund handset of those who purchased the phablet.

Second, the company will notice quickly about the competitor new strategy. With real-time analytic, the company will get notification about the rivals the moment they change their strategy, technique or lowering the price of their product. It can be done by "like" the rival page at Facebook, "follow" them on twitter, or other social networking. Ensure the company tracks both of the brands and what the rivals are doing. For example, as an owner of a restaurant, they may want to follow the rival restaurant name and chef in order to get the picture of their movement. Other methods are by connecting torival mailings and promotions, where the company signs up for membership to the rival company e-mail and direct marketing campaigns, normally known as customer relationship marketing programs. After becoming the regular customer, the company will rapidly notice what sort of offers the rival give and how they stack up to the company.

Third, rapidly improve in service, which brings to high changes and additional income. As the company observes item or product that is used by the customer which the product can proactively react to the breakdown in the future. For example, car sensor where it will show a picture of the engine in the display makes the driver realize that the car needs to bring to the workshop for a checkup. Fourth, it can minimize the damage to the company when the company uses the real-time analysis as it can detect suspicious activity when it happens such as fraud and the company can stop it or block it. Criminals usually targeting the financial market where any attempt of criminal to hack to the company or organization will be quickly notified by using the real-time tool and the security division try to block the hack while informing it to the authorities do deal the criminal.

Fifth is cost saving. Many companies think that if they use the analytics tool for real-time big data it might be costly; eventually it saves lot of money in the long run. The company and in memory databases can minimize the time of waiting and lessen the burden on the organization, freeing up asset previously reserved for reacting to demand reports when the company utilizes the real-time analysis. Sixth is excelling in sale insight which can bring extra income. Using the real-time analytic tool, it will inform how the company sale is doing and also inform the retailer which item or product sell well so that the company can do something to avoid being outdated or get a loss. Other real-time big data analytics advantages are getting a valuable data about the new and current customer trends from the competitor by keeping track what the competitor offer which can be in term of how the competitors promote their product or consumerism.

Cost: As a result in the advancement of big data and analytics, most business organization today are now tapping into investing more on latest big data and analytic technologies to cut operational cost for the organization. First, is in terms of cutting cost through storing the data. In the past, companies use to store the vast data in a place called data warehouse. It involves large and expensive hardwares that are capable to store a large amount of data in one location. In addition to that, additional cost may be involved in terms of having to upgrade the capacity of the infrastructure in order to be able to handle the ever-increasing amount of data collected daily.

Business organization may also incur additional cost as the large and sophisticated equipment needs to be placed at a secure location. Hence, typically they would store this equipment either by building a room or even rent a location to place them. Both of these actions will be costly for the organization.

As big data evolved and new technologies are now available, business organization are now able to cut or reduced their business operation by not having to buy expensive computer infrastructure and also not having to rent space to keep those often large and space consuming equipment. The development of today's big data technology is that it does not require a large computer as it has become more advanced in terms of storage capacity and reliability as well as accessibility and it only cost a fraction of what the old method cost the organization earlier.

Forexample, BI360 which is an update of traditional data warehouse that is predesigned data warehouse based on one of the most popular or leading Microsoft Structured Query Language (SQL) server platform. This latest version of data warehouse has the ability to integrate most of the business transactional data and placed it into a single BI database system in which one special feature is that this database can be managed by the company's business users.

Therefore, with this new version of data warehouse, all of the business vital information, whether it is in-house or even cloud-based, can be put away and is accessible for consolidated reporting, analyzing, and dashboards. The best thing about this new system is that it is relatively cheaper compare with the old method and so what this means is that businesses can reduce their cost.

Some organization that utilizes big data firmly believe that with modern technologies, such as Hadoop clusters, million instructions per second, and terabyte storage for structured data are now cheaper and affordable.

To illustrate their explanation of why it is more cheaper, for example, the cost of storing data through the traditional big data is approximately 1 TB for a year cost approximately \$37,000 and about \$5000 for a database appliance. In comparison, the new technology of big data that is Hadoop cluster now cost the company approximately \$2000 only. From this, we can see how cheaper the newer technologies have become when compared with their predecessors.

8.5.2 DRAWBACKS

Unfortunately, the development of big data does come with some negative implication as well. These implications may mean that the business might face other issues brought by the development of big data.

Too Much Data: The first drawback of big data paradigm is that there are too much or big information that can cause those who analyze them to lose track of direction of what the company initial intention in gathering data or the objective of the search.

A huge amount of data can make the analysis to become too complex and extremely complicated which could lead to misinterpretation of data analysis and results in information that is not accurate. As a result, the company will make a bad or wrong decision which makes the company face serious consequence such as the company will waste valuable resources and money.

Another issue that too much data can bring to the company is in terms of time. This can be true for small companies who have little resources such as good tools and software to analyze the huge data effectively. Or even perhaps do not have the adequate workers who possess the skills and knowledge to analyze huge amount of data.

Security: By using big data to store data, there is a risk that the information may get malicious malware which can cost the company huge loss due to the information that they have been gathering for many loss or the company strategies or secrets are leak or exposed to the public. The record shows that there is an increase of 23% of the identities report exposed from 2014, 348 million identities to 2015, 429 million identities which the larger number compromise comes from the health sector actually comprises 39% of all compromise in the year.

Forexample, a case involving security concerns that happened a few years ago, where Zeus Malware which was known for leaking bank information and as well as targeting the user's desktop which it could steal the user data. Another example, which involved point-of-sale (POS) malware which first appeared in 2013, where malware targeting payment processing systems such as using the credit card to purchase an item from the online shop or reserve a flight. The recent assault was last year 2015 showed that the malware was coordinated at 64-bits machine with the high version number. Another POS malware detected is Punkey, where the purpose was to extract personal information. The common victims of

POS malware are businesses such as shop, hotel, and restaurant that use the credit card as a payment.

Privacy Issues: As from what have been discussed regarding ethical issues of big data, one of the besttalks about drawbacks of big data is privacy issue. As the technology of big data evolved dramatically and become more complex and advanced its ability to abstract the personal data of targeted customers almost discreetly proves to be a critical issue brought upon the customers in retaliating the action of the business organization in doing so. Customers argue that they did not or are willing to give out their personal detail to these business organizations.

One of the probabilities that big data can bring upon an organization is, if they found out that these companies or business organizations get their personal information, they would either complain to the authority concerned or if they have a strong case could file a suit against the company.

In addition to the above, the company's reputation may be affected by all this, as in the eyes or the community or customers specifically, they violated their trust in the company and may now avoid or boycott the company altogether which could destroy their good brand name.

Can Be Costly: Another drawback of big data paradigm shift is in terms of the capability of storage increasing. The more advanced the data technology is becoming the more data size it can retrieve, which will ultimately mean more storage is required in terms of capacity of the hardware that is required.

This would mean organization or companies to invest more expensive and high-performanceIT equipment to be able to store and also to process the huge amount of data.

As an indirect impact of acquiring more advanced equipment and software, this will also mean companies or business organization will have to train current staff and managers to ensure they could run this BI and analytical process technology adequately. All this will mean the company will have to spend more to run these big data technology.

8.4 CONCLUSION

BI and analytics have become one of the most discussed subjects both in the business communities as well as the scientific research community specifically related to healthcare. The paradigm shift of big data has a profound impact toward the technological aspects of BI and analytic in where it has evolved so much that it creates more opportunities that organization can utilize.

From the report discussions, these technological advances have made it possible for organization to process huge amount of data that enables to make strategic decision accurately based on the data. More reliable data also was made possible due to this development as now big data has the capacity to analyzed these huge data through past records and predict certain behavior or consumers and answer questions such as what had happened, why it had happened and come up with some possible solution to overcome the problem. As from what have been discussed earlier above, one of the most amazing new capabilities of today's big data technology is that it can analyze and predict future outcome which would prove extremely helpful for top managers in terms of being able to make the right strategic decision.

Another element that makes big data quite successful for organization is in terms of the cost involved. The traditional method of big data cost more than what the current version cost. Not only does it cost much less, but also the amount of data is huge which could yield a much more reliable data information.

Unfortunately, the paradigm shift of big data technology also brings some challenges for business organization to deal with. In brief, one of the most significant challenges is in terms of finding suitable manager and workers to be able to use these modern big data technologies.

Conclusion based on all the discussions above, the evolution of big data has a profound effect toward today's business organization. The ability to make quick and precise decisions proved to be a key factor in most large business corporation strategy and big data can give them all this. Modern and up to date tool's has enabled management to process a huge amount of data and be able to retrieve it rapidly and importantly able to analyze the data and give meaningful information for the user.

8.5 RECOMMENDATION

Based on this report the company should already recognize that big data can help improve the business. First, log analytics is a basic establishment of numerous business big data application. Log management and analysis tools have been around much sooner than big data. However, with the exponential development of business activities and transaction, log data can become a huge pain to be stored, prepared, displayed in the most efficient, and cost-effective manner.

Numerous business and open source log analytics tool like Open Web Analytics can give the company the ability, to gather, to process, and to analyze enormous log data without throwing away the data into relational databases and recovering it through SQL queries. The cooperative between log search abilities and big data analytics has allowed theorganization to find insight for more agile operations. Nowadays, big data log analytic application are widely utilized for different business goals, from IT security system and network performance to market trends, and electronic commerce personalization.

Second, an organization that manages a large amount of financial transaction keeps looking for more innovative, effective ways to deal with fraud. Medical insurance agencies are no exemption, as fraud can cost the business up to $5 billion yearly. In the traditional method of detecting fraud, fraud specialists need to do work with BI analysts to run complex SQL queries from bill and claim information, then holdup weeks or months to get the results back. This procedure sometimes causes long delays in legal fraud cases thus, huge losses for the business.

With big data technologies, the investigator can analyze individual record by searching on a graphical interface since all the billing and the claim records were processed and stored inside the search engine. Predictive analytics and machine learning capabilities allow the fraud detection to give warning alert when it identifies a pattern that is similar to the previous fraud scheme.

Third, limit the threat tothe organization. The organizationeither gain proof or are informed by employees of the same organization who are able to steal company secret information, breaching organization, or perform actions that may be damaging to the whole organization. The company gets notified about what data is being lost and by whom, who among their employees is technically skilled enough in assaulting the company framework or system and who is showing strange behavior during useof system data.

Fourth, defense is against cyber threat and attack. The company is informed about what will happen or current assault by criminal enterprises or government funded colonies. By this way, the company will gain data about the source of the attack, the weapon that was used for hacking and person that utilize hacking, the sign of attacks assuming any so that the symptom can be utilized for foreseeing future attacks.

Fifth, Brilliant IT Investment and Rationalization. Many organization-sneed to recognize what their customer want from their website, which features that the customer use and do not use, how that behavior change when the customer is on their tablet or mobile device then suddenly the connection goes slow and many more. The new world of data gathered in context will permit the organization to quickly understand the organization customer value, need and want, likes and dislikes, behavior then utilize the data to target IT investment to the area that will have the greatest positive effect on their customer and on their income.

Sixth is big data and customer relationship management (CRM). With the big data, the company can collect data about the customer and analyze the data so that they see the better picture about the customer when they are facing operation also see the pattern about the behaviors of the customer. With the big data, the company can predict and decide return on investment and use it to embrace additional CRM investment.

It also improved customer analysis as it evaluate all customer touch point that includes social media such as Facebook and Instagram, e-mail, the Internet such as visit the company website, visit the company online retail store, give rating and review to the product, and call center which enables CRM and big data to segment customer to actions. Customer pattern can be mined from big data and used to predict needs, directing product improvement, and promotional efforts.

KEYWORDS

- **business intelligence**
- **data paradigm**
- **online**
- **real time**
- **warehouses**

REFERENCES

Azli, A. Samsung Note 7 Recall: What Should Owners Do. The Brunei Times, Sep 7, 2016. http://www.bt.com.bn/science-technology/2016/09/07/samsung-note-7-recall-what-should-owners-do.

Barlow, M. Real-Time Big Data Analytics: Emerging Architecture; O'ReillyMedia, Inc. 2013.

Boddy, D.; Boonstra, A.; Kennedy, G. *Managing Information Systems: Strategy and Organisation* (3rd ed.); Pearson Education Limited: England, UK, 2008.

Chang, A. C. Big Data in Medicine: The Upcoming Artificial Intelligence. *Prog. Pediatr. Cardiol.* **2016,** *43,* 91–94.

Chang, R. M.; Kauffman, R. J.; Kwon, Y. Understanding the Paradigm Shift to Computational Social Science in the Presence of Big Data. *Deci. Support Syst.* **2014,** *63,* 67–80.

Chen, M.; Mao, S.; Liu, Y. Big Data: A Survey. *Mobile Networks Appl.* **2014,** *19*(2), 171–209.

Chen, H.; Chiang, R. H.; Storey, V. C. Business Intelligence and Analytics: From Big Data to Big Impact. *MIS Q.* **2012,** *36*(4), 1165–1188.

Fayyad, U.; Piatetsky-Shapiro, G.; Smyth, P. From Data Mining to Knowledge Discovery in Databases. *AI Magazine* 1996, *17*(3), 37.

Gandomi, A.; Haider, M. Beyond the Hype: Big Data Concepts, Methods, and Analytics. *Int. J. Inf. Manage.* **2015,** *35*(2), 137–144.

Hashem, I. A. T.; Yaqoobl, I.; Anuar, N. B.; Mokhtar, S.; Gani, A.; Khan, S. U. The Rise of "Big Data" on Cloud Computing: Review and Open Research Issues. *Inf. Sys.* **2015,** *47,* 98–115.

Hemmatfar, M.; Salehi, M.; Bayat, M. Competitive Advantages and Strategic Information Systems. *Int. J. Bus. Manage.* **2010**, *5*(7), 158.

Lim, E. P.; Chen, H.; Chen, G. Business Intelligence and Analytics: Research Directions. *ACM Trans. Manag. Inf. Syst. (TMIS)* **2013,** *3*(4), 17.

Lohr, S. The Age of Big Data. *The New York Times,* Feb 11, 2012.

Madden, S. From Databases to Big Data. *IEEE Internet Comput.* **2012,** *16*(3) 4–6.

Mateosian, R. Ethics of Big Data. *IEEE Micro.* **2013,** *33*(2), 60–61.

Smith, N. C.; Goldstein, D. G.; Johnson, E. J. Choice Without Awareness: Ethical and Policy Implications of Defaults. *J. Public Policy Market.* **2013,** *32*(2), 159–172.

Susanto, H. Smart Mobile Device Emerging Technologies: an Enabler to Health Monitoring System. In *High-Performance Materials and Engineered Chemistry;* Torrens, F., Balköse, D., Thomas, S., Eds.; Apple Academic Press, 2018, pp 241–264.

Tahir, D. Predictive Analytics Play New Role in Fraud Detection, but Critics Want More. *Mod. Healthc.* **2015**. http://www.modernhealthcare.com/article/20150225/NEWS/150229947.

Wielki, J. Implementation of the Big Data Concept in Organizations – Possibilities, Impediments and Challenge, FedCSIS, 2013, 985–989.

Wise, S. City Worker Abused Authority for Personal Gain, Investigation Shows. http://wtvr.com/2016/07/13/dss-investigation (accessed July 13, 2016).

Yakabuski, K. Big Data Should Inspire Humility, Not Hype. The Globe and Mail, March 4, 2013, A11.

CHAPTER 9

CRAFTING STRATEGIES THROUGH ICT AS AN ENABLER OF BUSINESS PROCESS REENGINEERING

HERU SUSANTO

The Indonesian Institute of Sciences, Jakarta, Indonesia,
E-mail: heru.susanto@lipi.go.id

Tunghai University, Taichung, Taiwan

ASTRACT

The competition level has been increasing dramatically for business. Business process reengineering (BPR) provides a solution to this issue. Many corporations have become successful with BPR. BPR is the continuous modification and redesigning of business processes to achieve more improvement in quality, cost, performance, services, and response in a business, whereas business processing is the activities that implement what a consumer or the market desires for a product or service. Through IT business can become more valuable for consumers both internally and externally. IT allows an efficient and effective change in when work is performed. The successfulness of BPR depends on critical factors such as how to implement IT into process reengineering. IT is a major core of BPR and was called as a major cause for the change. Continued improvement in IT tends to create a high-executing organizational design while also contributing suppleness to redesign business processes to organizations.

9.1 INTRODUCTION

Modern IT has significantly improved and is able to perform more tasks and has more abilities compared to the older days. One of the examples is enterprise software which provides common data infrastructure within an organization and the ability to share and view information freely for their daily tasks (Ramirez et al., 2010). Broadband network also enables employees to freely share or transfer data in a short period of time regardless of the location. Better corporate decisions can be made through business intelligence applications as corporate data will be well analyzed and displayed during decision-making (Ramirez et al., 2010). Internet-based network enables the manager to control and monitor overseas branches and to make sure that the decision was carried out properly (Ramirez et al., 2010). This is very convenient and beneficial for centralized managers.

IT can be advantageous to business and is critical to the success of BPR; however, its effectiveness greatly depends on the user's behavior. End users must be willing and able to operate with the latest introduced software programs (Bondarouk and Ruel, 2008). IT also requires a huge investment in both IT equipment and labor IT courses. Losses can be great if IT is not fully utilized up to its maximum extent for it has the potential to develop. Despite the risks, rewards can be highly rewarding if IT was implemented in BPR. New forms and types of human resource management (HRM), decision authority, organizational structure as well as better application of decentralized decision-making authority, and the use of self-directed teams will be the result of implementation of IT in BPR (Ramirez et al., 2010).

9.2 LITERATURE SURVEY

"IT enabling of BPR in organizations" highlighted that BPR is a control approach in which organizational performance is improved by increasing the efficiency and effectiveness of operations by overhauling core services (Sungau et al., 2012). Its main idea is that although BPR enhances the efficiency of processes, it cannot be implemented on its own and, therefore, needs concepts from other fields. IT tools are perceived to be effective and accurate enablers of BPR approach in restructuring activities. Their paper focused on evaluating how IT can be an enabler of the BPR concept to restructure processes in an organization with the purpose of improving their effectiveness and efficiency. The report also presented the roles of IT in BPR for organizations and the challenges for them when implementing BPR.

Lee et al. (2009) claimed the significance of IT on BPR from an intra- and extra-organizational viewpoint. Their work proposed a foundation for promoting BPR attempts toward competitive businesses. The framework was tested from a survey on a sample of 382 senior information systems (IS) managers or chief information managers. The study results showed that competitive tension, organizational modernization, and market pressure emphatically influence IT adoption, thus causing adjustments in the business process in terms of business structure, working environment, and personnels.

Exploring the relationship between IT competence and quality management have pointed out that the introduction of IT has become essential to compete in most markets; so, simple implementation of an IT scheme is not enough to attain a better firm performance (Pérez-Aróstegui et al., 2012). They reviewed literature work done by others and concluded that IT is an effective tool only when it is complemented by other assets and practices. The purpose of their research was to investigate the relationship between IT and quality management practices (QMP), one of the most successful and common sets of the organizational system.

Chan (2000) stated that countless forms of information technology (IT) have a visible and complex influence on consumers of IT and their environments. This paper proposed and explained a structure of the roles of IT as an enabler, an initiator, or a facilitator. It also demonstrated the application of this framework by an analysis of the impacts of IT. Further, Chan examined the growth of IT's repercussions on business processes through definitive type of technology—specifically the imaging technologies, computing, and telecommunication. Last, he reviewed IT's influence on management and institutions and gave a suggestion that an extensive and comprehensive understanding of the functions of IT will also enable systematic identification and evaluation of the costs and risks associated with relationships concerned with implementing IT in business processes in an organization (Susanto, 2018; Susanto et al., 2018; Almunawar et al., 2018b; 2015a; 2015b).

Willcocks and Smith (1995) in their paper titled "IT-enabled BPR: organizational and human resource dimensions," addressed IT-enabled BPR that can be brought to organizations. First, it inspected the relevant literature and a current United Kingdom (UK) survey and suggested that BPR activity was too often methods-driven. Those methods tend to be limited in their approach to problems that required a more holistic outlook.

Particularly, the paper stated that IT-enabled or IT-driven BPR programs presumably diminish the attention to social, political, and human processes, ignoring the fact that these are strong elements of success or failure. Willcocks and Smith utilized three case studies from health care, aerospace, and pharmaceuticals to pursue these determinants. They expressed that BPR caused political problems that are intrinsic and important to BPR activity. The political aspects of BPR are examined and a proposition to handle the organizational and political human issues was also developed in the paper.

John Qi Dong, Chia-Han Yang in their study entitled "Information technology and organizational learning in knowledge alliance and networks: Evidence from U. S. pharmaceutical industry" in the year 2014, acknowledged that in the recent years, there has been a change in businesses' innovation style, from open to closed, of which IT has been an integral part. This study focused to open up the black box of IT-enabled absorptive capacity by hypothesizing and experimenting with the role of IT in two organizational learning processes, which are either nonreciprocal with others in the knowledge chain or reciprocal with partners in the knowledge alliance. Specifically, John and Yang formulated a model which explains how a company's IT expenditure alleviates its organizational learning processes in knowledge networks and affiliation, which provided enlightenment on the parts of IT as an enabler. They also used a longitudinal data set from the US pharmaceutical industry to illustrate the results of IT's role in enhancing the organizational learning processes which improves the business competitiveness.

By undertaking a study on three cases, Bondarouk and Ruël (2008) in their work, "HRM systems for successful information technology implementation: evidence from three case studies," discovered that the benefits of IT projects are extremely reliant on the end users' practice. The paper stated that the willingness and ability of users to operate with the newly introduced software applications is essential and, therefore, a crucial concern is supporting the targeted workers of freshly introduced software applications in their appropriate utilization. Bondarouk and Ruël believe that HRM practices have the capability to aid in the endeavor. This article expands on the HRM systems for software employment fixated on three HRM areas: supplying chances to work with new IT, reducing difficulties in its use and making sure staffs are able to use it. The paper specified 17 HRM behaviors that should be included in the plans of IT programs if they are to accomplish a devoted and proper usage of newly-introduced IT by the selected workers based on findings from 83 interviews carried out in the three case studies.

A journal, "Role of Human Resources Management in the Effectiveness of Business Process Reengineering" written by Naz et al. (2013) focused on presenting a model with the analysis of human resources management role with the line extent of effective communication, teamwork, management proficiency, IT and organizational structure, and their influence on the performance of BPR. The paper also suggested that the implementation of HRM with the usage of the BPR achieves the better outcome and assists in improving the performance of the organization.

Ruiz-Mercader et al. (2006), in their empirical study named "Information technology and learning: their relationship and impact on organizational performance in small businesses," looked at IT as a major tool in knowledge management processes. The paper, however, stated that the existence of IT neither assures knowledge use, knowledge creation nor knowledge distribution. In addition to IT, a working space and culture which promotes steady learning and sharing should also be initiated and upheld by the authority. This paper also provided empirical evidence of the link between IT and learning in small firms along with its importance on organizational performance. The findings revealed that individual learning and cooperative ITs have a notable and positive influence on organizational learning. On the other hand, individual and organizational learning have shown positive and consequential impacts on organizational performance unlike individual and collaborative IT. The researchers then concluded that IT has a significant effect on results only when a proper education situation is in place (Susanto et al., 2018; Susanto and Chen, 2018a; 2018b Susanto, 2017a; 2017b; 2017c).

Jan vomBrocke, Alessio Maria Braccini, Christian Sonnenberg, Paolo Spagnoletti (2014) in their work "Living IT infrastructures—an ontology-based approach to aligning IT infrastructure capacity and business needs," investigated the importance of ontology for linking processes and IT infrastructure (OLPIT) in the relationship between both IT infrastructure and business process activities. Through the results of their interviews, they suggested that OLPIT is advantageous when it is integrated into the accounting information system compared to programming a system that can only operate independently. From the limitations of the prototype used, they realized that using OLPIT to operate businesses can help software supports to be more efficient, for example, Excel enables teamwork. Their extended research also showed that integrating both configuration management databases (CMDB) with the accounting information system is also prospective as it provides storage space for all information related to IT components and allows integration of data sources. The ontology was created and analyzed in the condition of a research that was designed by Hilti Corporation, which explains the effect of business and service costs on IT.

The paper titled "IT enabled Business Process Reengineering" by Madhumita Panda (2013) defines BPR as "the fundamental rethinking and radical design of business process to achieve dramatic improvements in critical, contemporary measures of performance." The study examined the impact of IT in BPR, how IT development helps to promote options for process execution, and how it opens up chances for reengineering to take place. The paper also included the aims of BPR which is to make businesses more efficient and to reduce costs. A case study was carried out based on Hindalco that showed the successful implementation of enterprise resource planning system that brings different benefits to an organization (Almunawar et al., 2013a; 2013b). The paper concluded that IT provides skills and tools needed for effective reengineering and therefore, without IT support, reengineering is impossible.

According to the study, "The role of information technology in strategic buyer-supplier relationships" conducted by Hannu Makkonen and Vuori Mervi (2014), IT is defined as an element of structural layer in relationship management which supports the action layer by encouraging changes and maintaining relationship stability. Case studies were given to show that IT acts as a communication forum that manages the organization's relationship by creating opportunities for operation value creation and effective joint operations between organizations. This helps to facilitate the implementation of both the buyer's and supplier's objectives in creating a framework for effective performance. They also conducted further research on both quantitative and qualitative studies that show how IT is linked to the structure and various activities within an organization in the buyer-supplier relationship.

In their work entitled "Strategizing IS-enabled organizational transformation: A trans-disciplinary review and new directions" by Patrick Besson and Frantz Rowe (2012), they pointed out the problem on how information system enables organizational transformation. They examined the discourse related to organizational transformation strategy, theory and information literatures, and came up with four structuring themes that consist of organizational performance, inertia, process, and agency. Sixty two empirical papers were found related to these themes and the results were discussed. The results provided a clear understanding on how IT is related to organizational transformation and the effects of process reengineering on organizational inertia. Ten avenues were identified and results concluded that organizational transformation is still a new frontier for strategic information system research.

In "The Role of IT in Business Process Reengineering," G. P. Sudhakar (2010) discussed about the importance of top management's commitment, skilled reengineering team as well as communication in successful implementation of the BPR process. Sudhakar noted that IT presents significant advantages when combined with BPR. Notable pros include reducing turnaround time, more accuracy and precision, improved communication, increased productivity and efficient progress tracking. Adequate staff training will be required in order to reap the benefits of BPR. Automated BPR implementations are recommended for the application to complicated processes or precise calculations such as payroll calculations. It was concluded that the combination of IT and BPR would benefit the major stakeholders of an organization.

A. Gunasekaran, B. Nath (1997) examined the role of IT in BPR, particularly in manufacturing industries in their paper entitled "The role of information technology in business process reengineering." Business processes such as operation and production segment remain the artery of the industry; consequently, process simplification is the principal step in BPR efforts. The paper presented a conceptual model identifying various possible applications of IT in major business processes of an organization. The business processes identified are namely, product design andengineering, order processing, marketing and sales, personnel management, accounting, service, strategic process and technology. The authors presented their suggestion on how IT will be able to play a major role in eliminating barriers between various business process and functional areas. For instance, the distribution (order processing) function can be incorporated with Internet and online inventory and shipment controls, database, barcoding, distribution resource planning (DRP), satellite positioning, electronic data interchange (EDI), and custom clearance. This integration will eliminate processing and communication barriers with marketing and sales, purchase and production functions to facilitate a shorter lead time for goods production and delivery to customer. Similar to Sudhakar's work, this paper also implied the importance of top management support and a skilled reengineering team to focus on the formulation and designing to productive implementation of BPR. Analysis must be conducted for target objectives to accelerate effective BPR. Furthermore, organizational change can be brought about by BPR initiatives in cases where basic knowledge and culture in the organization has to be redesigned, re-instilled, and reinforced. Top management support is crucial at this point, where motivation and incentives in the form of rewards or monetary incentives can be awarded to staff in the event of a successful organizational change. Gunasekaran and Nath concluded that the integration of IT and BPR in the manufacturing industry ultimately improves the productivity of the organization and efficiency, and it is believed that other industries such as the service industry should be able to integrate and implement IT and BPR for greater business productivity and quality.

Ronald Ramirez, Nigel Melville and Edward Lawler (2010) in their analysis named "Information technology infrastructure, organizational process redesign, and business value: An empirical analysis," employed analytical research into the BPR portfolio data sets created by matching firm-year observations from three secondary datasets. Successful implementation of BPR initiatives increases an organization's BPR portfolio and it was noted that the sample consisted mainly of conventional, industry era firms. The authors established that BPR in an organizational context affects all internal segments including human resource, customer service, production and so on, and therefore, would be considered extensively by the top management. A model framework was used to perform variable analysis between different BPR and subsequently, market value estimation. The authors concluded that the interaction between an organization's IT and process redesign portfolio is positively and significantly associated with the organization's production efficiency and market value. The conclusion was made in assumption that the management in organizations applied adequate and sufficiently adaptable BPR methods. Process improvement may be evident immediately after the implementation of BPR initiatives, yet the authors suggested that the management should adopt a more conventional approach in organizational change efforts if the objective is for long-term market impacts. In other words, gradual improvement or the "kaizen" approach for BPR integration with information technology is recommended for effective and sustainable organizational change.

Sangjae Lee, HyunchulAhn (2008) in their work of "Assessment of process improvement from organizational change," stated that even though BPR has shown to improve organizational productivity which in turn leads to profitability, the costs involved in initiating and maintaining BPR in business processes remains one of the major factors preventing businesses from investing in BPR. For multinational companies, the investment sum for a large-scale BPR in operations may involve millions of dollars. In this case, the return of investment (ROI) of BPR-related cost of investment is required for top management to come to a decision on whether the implementation and maintenance of BPR is feasible. The researchers provided the management with an executable model process improvement using four analysis tools—specifically Task Activity, Bottleneck, Cycle Cost and Resource Utilization. Through data inputs and calculations comparing different reengineering alternatives, quantitative business process improvement can be derived in the form of time and cost savings as well as margins for error. The authors believed that by applying the four analysis tools, organizations would be able to conduct impact analysis and make well-informed decisions when contemplating BPR initiatives.

9.3 REENGINEERING EMERGING TECHNOLOGIES

Reengineering which is also termed as redesign or constant enhancement of business processes is a necessary process to keep up with numerous and growing competitors either from public or private organization. It is also needed to endure a constantly changing environment and technological factors (Tsalgatidou, 2014).

Tsalgatidou (2014) also mentioned several methodologies of BPR such as those introduced by Hammer, Champy, Davenport and Short. An organization can decide on using a different BPR method, depending on their targeted success factors for example, an organization's vision, subordinate employee value, customer satisfaction, financial performance which have been identified by the organization's top management. Therefore, understanding the need of the business change is important before the BPR project teams decide to implement the preferred BPR approach.

The old work processes which had occurred before the beginning of Internet communication and technology was when the challenge is massive in implementing the BPR project. The process of reengineering could involve several linked aspects such as business processes, values, beliefs, and job structures. There are six phases of the BPR implementation introduced by Hammer and Champy:

1. Introduction into business reengineering
2. Identification of business processes
3. Selection of business processes
4. Understanding of selected business processes
5. Redesign of the selected business processes
6. Implementation of redesign business process

When compared to Hammer methodology, Davenport and Short's BPR methodology emphasizes on three important aspects such as the technologies, processes, and human resource during the implementation of BPR (Mohapatra, 2013). As stated by Homeas H. Devenpoirt (as cited in Mudiraj, 2014, p. 1), BPR has embraced foreseeing of innovative work strategies and the actual process design action, which executes changes on complex technological, human, and organizational diminution issues. Also, it helps an organization to be productive and manage effective decision-making by redesigning the business process according to their goals and milestones. The implementation of BPR has similar processes to the traditional lifecycle of the software development which consists five phases, namely, the planning for BPR, data collection and analysis, designing the BPR process, implementing the BPR process, and lastly, testing BPR progress.

According to Muniraj (2014), the planning of BPR process trails after completing the enterprise resource planning (ERP) during classification of the current business gaps and processes. The organizational top level management has to be optimistic and ready with the BPR processes, so that they can come up with a better mission, vision, and improve their existing business objectives. During the data collection and analysis phase, the organization will start to gather information from both external and internal resources. This data gathering includes high authority's mission and vision, protocol of the organization, existing technology, difficulties and risks encountered during the current business process, and constraint issues on cost and time. Therefore, the impact level of risks on the current processes needs to be identified before proceeding with the designing of the BPR process.

During the designing phase, the organization will be prepared with detailed information on the issues and methods of old business processes. The BPR teams will come up with a step-by-step approach and distribute the entire task amongst the team members. Consequently, these BPR team members will obtain possible and positive solutions to the targeted problems. Next will be the implementation phase, where reengineering faces the greatest resistance. The implementation of BPR process has two basic strategies, implementing BPR completely or partially. During this phase, the observation of structural and behavioral changes in the process as well as people who are involved directly or indirectly will be constantly monitored. This is followed by the testing phase, where benchmarking of the tests will be recorded to classify whether the new business process can be achieved or not. The continuous monitoring is required in order to analyze the outcomes whether they are acceptable or the organization should proceed with the alternative solutions that they have identified during the implementation phase. Thus, the process will be repeated until their new objectives are achieved (Muniraj, 2014). This method was also affirmed by Doumeingts and Browne (1999).

9.4 MANAGING CHANGE

9.4.1 ISSUES WITH THE IMPLEMENTATION OF BPR

There are several major issues which must be considered and/or resolved before or during the implementation process of BPR. Failure to do so may cause problems for the organizational process both in the short and long run. Undoubtedly, the failure will have serious impact for the company and all those involved including the company's employees and stakeholders. A number of issues will be discussed hereon.:

9.4.1.1 ECONOMIC FEASIBILITY

BPR requires a significant financial investment during the implementation process for example with the purchase of hardware and software, employees' skills development, and also the hiring of technicians and consultants (Jurisch et al., 2012). The addition of resources will increase the cost as well as running new hardware, maintenance of the new information system, and also upgrading the building's infrastructure to run the system (Jurisch et al., 2012). Top management would need to determine whether or not the implementation of BPR would be worth the investment to generate significant ROI.

9.4.1.2 EMPLOYEE PERCEPTION

As stated by Mohapatra (2013), most employees believe that BPR is a fancy word for downsizing, which is used by the top management as an excuse to lay off people, thus making the employees' behavior rigid and resistant to change. Employees may also have encountered difficulty in dealing with day-to-day tasks where unfamiliarity may cause complications and hiccups especially during the BPR implementation. Organizational restructuring could leave some employees dissatisfied if they are suddenly passed over or demoted. Some employees who fear the unknown changes could contribute to resistance to change (Mohapatra, 2013).

9.4.1.3 ORGANIZATION RESISTANCE TO CHANGE

Organizational issues are the nontechnical aspects of system development (Violet andWatundu, 2011). Change is also necessary to maintain a competitive advantage, but it may not be the easiest process. Managing individual resistance is easier than organizational resistance because a group of people who have worked together for years and have developed a sense of cohesiveness may cause organizational inertia (Violet and Watundu, 2011).

9.4.1.4 INFLEXIBLE ATTITUDE

Flexibility, adaptability, and an open mind are essential in the development team as well as the management and employees. Problems may arise if the management and employees are unwilling to compromise with the BPR development team in order to achieve the desired goals set by the top management (Mohapatra, 2013).

9.4.1.5 TIME REQUIRED TO DEVELOP BPR

Software development, testing, and troubleshooting are essential during the implementation period. The system should be established before being deployed, taking into account the most important requirements of the implementation (Nah et al., 2001). The overall implementation period may require years, before being able to be used, depending on the complexity of the system being produced. This may not be ideal for an organization which prefers immediate and effective solution.

9.4.1.6 AVAILABILITY AND LITERACY OF IT

Huang and Palvia (2001) stated that BPR helps an organization to improve efficiency and effectiveness of their operations between the suppliers, consumers, and other external stakeholders. The maturity of IT also plays a significant role in a company's strategic planning in performing BPR as an IT-mature company can better collaborate with the BPR development team and hence, more likely to succeed.

9.4.2 BPR IMPLEMENTATION

The success or failure of BPR is dependable on a number of factors. These factors generally revolve around the users (or employees) of the system, the cost of the BPR or the organizational infrastructure. The following content will discuss a number of major factors involved in this implementation:

9.4.2.1 MANAGERIAL FACTORS BPR

In general, every proposal for change will come from the management. It is the management who initiates the changes and in this case, the requirements as deemed fit by BPR. These changes and BPR are requirements for the purpose of development and enhancement of the company to enable them to compete and place themselves as niche in their respective industry. It is the management's role to initiate and implement BPR. In efforts to ensure that every resource for the purpose of BPR's implementation is fully utilized and managed effectively, there will be no compromise on the effectiveness of results and the achievement of BPR. Hence, in order to achieve this, managers must have a clear understanding and knowledge on what needs to be reengineered so as to accomplish favorable results to achieve the objective/s of any changes that are designed for the company's road to success.

Furthermore, it is important for the management to have knowledge of the plans and objectives for the projects laid out for BPR, the direction that the team needs to go in, and at the same time to give their full support and encouragement to the BPR team so as to ensure its success. BPR development team requires the support and trust of the company's management to properly implement BPR. The management must be well-prepared, continuing on giving support and accepted delays on the resources and time. They should also set realistic expectations and avoid ambitious expectations during the initial implementation (Habib, 2013). The support of the top management indicates that the company is ready for change (Mudiraj, 2014). Other than receiving support and trust from the management, the BPR development team will also be required to have an open communication between the employees, the development team, and the company's management (Habib, 2013). The expectations and strategic goals at every level are also needed to be clearly communicated (Nah et al., 2001).

9.4.2.2 FINANCIAL FACTORS

Mudiraj (2014) had deliberated that a substantial initial investment is required for the purchase of hardware and software as well as for the hiring of technicians and consultants in order to set up the new system. An allocation of budget must be prepared for running the cost and maintenance of the new system. Employees would need to be well-trained in order to ensure the system is being utilized to its fullest. Other than the initial purchase price of the hardware and software, companies must also consider the cost of upgrades to establish and implement the new BPR processes such as the buildings, electrical, and Internet connection. The organization must prepare a contingency allowance (contingency reserve) to account for cost uncertainty. This is also viewed as miscellaneous costs to address the "known-unknowns" that can affect the BPR project (Project Management Institute, 2013). Other factors may be caused by the fluctuations in the stock market, currency exchange rates, and government policies.

9.4.2.3 TECHNICAL OR INFORMATION TECHNOLOGY (IT) FACTORS

Hardware being used to implement BPR must be up-to-date and compatible with each other. The outdated hardware may not have the capability or compatibility to implement, monitor, and control the BPR processes

(Mudiraj, 2014). The IT staff must be highly trained and fluent with the system to be effective in case of a failure (Mudiraj, 2014). IT is both the enabler and facilitator of changes (Trkman, 2010).Thus, in every single stage of the implementation of BPR, IT should be proposed and utilized. IT affects all processes that should be carried out by the company as well as the stakeholders involved such as, the suppliers and customers (Alghamdi et al., 2014).

9.4.2.4 OPERATIONAL FACTORS

During the implementation of the BPR process, it is important to have a better control over the operational activity (Wanare and Mudiraj, 2014). This would make it easier for the identification and analysis of different constraints such as time and resources (Wanare and Mudiraj, 2014). Risk analysis also plays a vital role in the BPR process, in which we can identify the internal and external risks which can harm the BPR process (Wanare and Mudiraj, 2014).

9.4.3 BPR IMPACTS

Both private and public sectors have the same goals of wanting to enhance efficiency and policy effectiveness in their businesses with the aid of knowledge in information system. However, there are few major differences. In terms of "value" or "quality", public sector is left behind a bit as this value is determined by stakeholders or the professional association of accreditees. However, in the private sector which always aims for customer needs, it is driven by the customer (Jurisch et al., 2013). Customers and their satisfaction have more say in businesses as they determine how the business would do. According to Jurisch et al. (2013), instead of achieving 47% of customer satisfaction, they achieved only 28%. As for the private sector, since their services are based on the customer satisfaction, they achieve 34% of 35% of their goal.

In terms of rules and regulations, public sector is bound with legal rules to obey and it has a restricted budget to reengineer its business processes which makes it unable to achieve its goal. In addition, public sector frequently changes its organizational chart so the business process is inconsistent.

Private sector also has to deal with new challenges everyday such as new competitors of the same type of business. Therefore, it can easily adapt into the situation where it can manage to accept its changing processes. On the contrary, public sector is supported by the country but it will also encounter some new challenges as well. Hence, when facing these challenges, some refuse to change due to the uncertainty in future which might lead to job loss and authority loss (Abdolvand et al., 2008).

In terms of cooperation, the private sectors can collaborate more with other organizations whereas the public sector is a little bit reluctant to do so because it refuses to share its data or findings as it is tied to rules and regulations set by the organization or the government. However, there is a possibility that the public sector will share its data if it is guaranteed with an efficient IS and confidentiality. According to Drake (as cited Jurisch et al., 2013), 46% of the private sector had collaborated with other parties compared to the public sector, whose collaboration was 39%.

9.5 BUSINESS CONTINUITY

The BPR methodology starts by defining the methodologies for the implementation of BPR and study the goals and objectives as well as the benefits of BPR. A list of research has been done regarding the issues and problems during the implementation of BPR. Several journals and research papers were studied in order to understand all this discussion. Other than focusing on the implementation of BPR for IS, this report also investigates and aims at perception of actual changes in the organizational business process with the approach of IS and the effects of BPR on human, process, quality, product, and other factors. Moreover, further research has been made in this report on the two BPR methodologies, issues and factors involved in BPR, and a list of similarities and differences in implementing BPR between the public and private sectors. Additionally, a few case studies were included in this study to learn more on the BPR achievements and also the success rates of implementing BPR.

A practical research was used based on this topic through a variety of documentation such as books, reports, magazine, websites, and most importantly, journals which can be accessed via the UBD e-library portal and downloaded from several databases such as Google scholar, Springer, Taylor and Francis, and Emeralds. This research is supported by at least 14 academic journals.

In order to highlight the benefits and impacts of implementation of BPR for IS, we have conducted two interviews by selecting one employee who is involved in the implementation of BPR for IS as the BPR team member and another employee who is involved as a user. This qualitative method is used to support our research finding and tailor it with a real-life case study. These participants will represent their personal point of view and how they define the implementation of BPR for IS.

However, the limitation of this research was the restricted choice of accessible journals. In addition, the good and latest journals need to be purchased or rented for viewing. Another limitation is the lack of time to conduct our own survey to analyze and support our findings on the result. A better result and discussion can be carried out if we perform the quantitative methods of data collection to expand our sample size. Ideally, at least 100 participants would have been sufficient in order to construct an even distribution of gender, a narrow range of age, work duration and the level of work experience of those who are involved in the implementation of BPR. As a result, this research survey would also expand our knowledge and increase our understanding on the chosen topic.

The presence of IS around the 1960s was commonly termed as management information systems (MIS) and the definition has been continuously progressing until now. The role of IS in the 21st century has been seen as a necessity for an organization to collect, process, store, disseminate information, and provide feedback in order to improve the organization's performance, productivity, and efficiency. Thus, they can provide better results and services to their stakeholders. Currently, most organizations are anxious in adapting their new business processes because of emerging issues such as tenacious technological, political, organizational, and enterprise environmental factors.

The success of IS is believed to be correlated with BPR which involves radical changes in structures and in processes within the business atmosphere. The possibilities of technological, human, and organizational structures are boundless in BPR. It has been widely implemented in organizations' business processes due to the number of success stories where the increase in productivity and efficiency are at their expected results. According to Grau, Franch and Maiden (as cited by Anand et al., 2013), BPR began in the 1990s, when Michael Hammer published an article in the Harvard Business Review called "Reengineering work:

don't automate, obliterate". Anand et al. (2013) also mentioned that the success stories such as the improvement in insurance writing efficiency under the Mutual Benefit Life Insurance Company had increased by 40% or the redesigning of order fulfillment process and improvement in service of Xerox had increased from 75% to 97%, were selected as an exemplar and baseline.

However, the definition of BPR has also been evolving through these years of implementation. Originally, Hammer and Champy in 1993 defined BPR as an achievement from the essential reconsideration and radical redesign to improve the existing organizational process which is correlated to the cost, service and boost the speed of the organization's performance (Zigiaris, 2000). Then, Hammer and Champy formally introduced the six phases of BPR methodology. Similar to his methods, Davenport, as an early exponent of this ideology, had presented the five steps methodology of the BPR model (as cited Mohapatra, 2013).

BPR is not just about taking a new approach towards taking different actions, but is also heading towards a new direction of development based on the latest IT which is crucial for any benefits that it may bring to an organization. According to Khuzaimah (as cited by Setegn, 2013), BPR could drastically affect every aspect of business nowadays as it can range from success to failure. Therefore, BPR could be a method that can be implemented to identify risks and point out the organization's business processes that need to be changed in order to obtain drastic performance improvement.

Lotfollah, Ziaul, Seyed, and Saeedreza also pointed out the primary objective of reengineering, which is to minimize wastage, enhance efficiency, and eventually decrease costs (as cited by Setegn, 2013, p. 116). Graham (2010) also stated that process reengineering is a useful tool to drastically help improve organizations. Through the implementation of BPR, an organization could increase customer satisfaction and improve their capabilities with better products and services. With the help of IS and BPR, the organization can be more flexible in meeting the rapid market demands which, according to Farmer (as cited by Jamali et al., 2011), would further improve the overall customer satisfaction, productivity, flexibility, employees commitment, workforce coordination, and create a new competitive advantage as a result of successfully implementing the BPR.

9.5.1 INFORMATION TECHNOLOGY (IT) GOVERNANCE

IT governance is a vital managerial skill to encourage critical IT-business alignment and IT value transfer to business. On applying IT governance, firms will have the opportunities to make use of specified practices concerning decision-making systems, processes, and relevant structures. However, the precise contributions of these practices continue to be poorly understood. (Bermejo et al., 2013). This case study aims to recognize the IT governance practices of successful businesses as well as the drawbacks of companies with lower business and IT results.

A study conducted by Information Systems Audit and Control Association (ISACA) with business and IT experts in Latin America identified a number of core IT governance difficulties such as high costs, low returns, and aggregation value of IT investments. In addition, this case study was based on quantitative data related to the maturity of IT governance practices and results achieved by the IT and organizations. IT governance practices were attained from the test conveyed by De Haes and Van Grembergen.

The survey obtained a total of 652 samples from Brazilian companies. 470 were private companies, 146 were public, and 36 companies were mixed (public and private). Regarding the employees, 317 companies have more than 500 workers, 52 companies have 250–499 workers, 134 companies have 50–249 workers, 44 companies have 20–49 workers, 65 companies have 5–19 workers and 30 companies have employees not more than five. A Structured questionnaire was handed to employees in managerial positions. In the questionnaire, the samples analyze the maturity of the IT governance practices as well as the results.

IT maturity was measured and evaluated using a six-point scale. First, nonexistent, where the practice is not conducted. Second, ad hoc, where practice is conducted occasionally. Third, intuitive and repetitive: the practice is repetitively conducted but not documented formally; fourth,) implemented and documented where the practice is repetitively conducted and documented; fifth, measured where the practice is measured for its performance and finally sixth, optimized and continuous improvement where the practice is fully conducted, measured, and improved by time.

9.5.2 PROJECT MANAGEMENT

Project management gets more complicated as changes of increasing speed may affect the business environment. IT has a higher influence on social and economic systems due to the major changes in business life. Changes are the basic reason to have more effective project management in enterprises. It is also suggested that today, software project management is an art (Boehm and Ross, 1989, 1). In this case, the importance of project management is increasing day by day, and the use of project management in most areas of IT is becoming widespread. On the other hand, a higher rate of failures in projects increases the need to understand the reasons behind the conditions for being successful in projects and IT project management. When the conditions for being successful are understood, projects can be completed as planned before. In this study, first of all, the definitions of success of IT projects as well as the factors are discussed. By conducting a comparative study, the factors leading to project success are investigated. In order to determine these factors in Turkey, a questionnaire is prepared and responses from 68 project managers are evaluated. The factors that affect the success of the project are analyzed by the logistic regression method. The results are compared with the ones obtained for Sweden and Australia, which is realized by Svensson (2006).

A project is a temporary endeavor undertaken to create unique products or services. It is a one time and problem-specific process that may secure the goal or target. The project and the project management in existence of a defined goal with specific characteristics or pre-defined beginning and final dates is a specific budget. They need minimum features specifying the usage of sources in order to rethink new method design products in extension strategies. In addition, there are some common features seen in the project such as complexity, originality, and ambiguity (Slack et al., 1998: 590). It is applying the information, ability, tools, and instrument to project activities to satisfy the consumer expectations and taste. Project management is measured by the degree of obtaining goals for three constraints and consumer satisfaction in order to be successful (Susanto and Almunawar, 2018; 2016; 2015; Susanto et al., 2011).

However, it is not easy to answer the questions, what are successful information technologies or how it is defined. In BPR, IT can vary according to the position in the organization and for instance, it is found successful for software developers. Mostly, a project which satisfies budgeting expectations and has the needed characteristics due to business goals and objectives is accepted (Susanto 2016a; 2016b).

A research study about the factors of success in IT projects, conducted by Standish Group, indicates that 61.5% of the projects in large companies may have exceeded the planned budget or may lead to "over-budgeting." The average increase in budget is 189% while the average schedule delay is 222%. It demonstrates that 60% of the software projects have reliability and quality problems. "The success of projects increased in 2000 compared to 1998 results and it rose to 28%. Examining the cost overruns reveals that the overruns were 189% in 1994 and 45% in 2000. While only 61% of the initially specified features of the project were covered in 1994, this ratio increased to 67% in 2000. This rise suggests that software industries have developed their abilities to realize successful projects. The dispersion of successful, unsuccessful and postponed projects can be seen in Figure 1 between the years 2000–2009" (Standish Group, 2009). Three important reasons that increase project management are smaller application parts, better management, and the usage of standards methods.

IT will improve and enable human resource dimension in many ways, for example, communication. A good communication system is necessary in many operations, for example, management competence. Management competence is the ability to reach organizational objectives, manage and use resources efficiently, provide excellent service to customers, and ensure high labor productivity. It is also a main part of HRM that helps to boost the competencies of staffs of the entire organization. Further workforce will prompt to improve and strengthen its abilities when the human resource team is efficient. Therefore, the human resource team is significant to make BPR more adequate. On the other hand, organizations will be at stake if obstructions are not removed, leading to BPR's failure. The top management is responsible for "communication-free" environment and communication software is installed with the help of IT so that most of workers in the organizations are free to communicate with each other to reduce barriers and lead BPR to become effective (Bondarouk and Ruel, 2008). This way of communication enables different departments in the organization to voice out their opinions and queries about where reliable information can be obtained and reduces the level of reluctance.

Organization structure falls under the responsibility of human resource for supervision, rules and regulations, task allocation, workforce environment, and other activities in the organization. For BPR to be successful, hierarchy system is implemented as the flow of information depends on the organization structure that leads towards communication between all the members in the organization (Naz et al., 2013). This system usually comes in the form of networking or team to establish the pattern of flexibility within all the team members and they must be willing to accept changes.

The results in Ramirez, Melville and Lawler's research shows that IT is also one of the key figures to business process reengineering. For example, as a manager, process change may not be able to be done without IT. However, the combination of IT and process reengineering as a dormant process is carried out for positive corporate change. (Ramirez et al., 2010)

BPR pioneering has become a beacon for managers who also discovered the importance of IT through the types of BPR projects selected. Moreover, communication menses are significant for an acknowledgment of business process reengineering with a proper design, where it can be seen that IT is an elemental for the business process reengineering project. The authoritative point of convergence of the research shows that the process determines the economical platform as one sense of means for organizationalrequirements. (Ramirez et al., 2010)

It shows that the organization would benefit from implementing BPR with the use of IT. The combination of IT and BPR gives a positive return in investment firms, both in terms of the value added and the firm market value. In any of the organizations, the number of process reengineering efforts commenced is enabled by IT. Managers have the opportunity in choosing the right and appropriate IT that enables their ability in various types and levels of BPR in the organization. (Ramirez et al., 2010)

IT contributes an efficient BPR to organizations which provide a high-transact organizational outcome. The information processing competence is certified by modern IT. For example, enterprise software provides a probable data support to an organization and also work cells with the opportunity of being able to do their work efficiently and productively. Moreover, broadband networks provide a convenient way that allows employee teams to get access among each other wherever they are through the network. Managers are able to keep track of the performance of their employees through Internet-based networks which enable them to follow their plan with all the implemented decisions and rules (Ramirez et al., 2010).

Business organizations typically make large financial investments in IT, often assuming that acquisition of IT is synonymous with the correct IT usage or that system integration is automatically in place (Sanders,

2008) Research also shows that IT plays an important role in process reengineering. The information competence by IT resulted in technology investment, as an important achievement in working methods with organizational change. Firms with higher levels of IT investment have been found to have a greater application of decentralized decision authority, use of self-managed teams, and cross-functional units (Ramirez et al., 2010) Influence of the communication knowledge and mechanized efficiency of technology provides a new form and type of organizational structure, decision authority and HRM. For example, a team may be more productive through the combination with a technology as IT to fulfill the information management and acquaintance among its members. Other organizational factors found to complement IT are employee behavior, worker composition, size, and culture (Liu et al., 2018; Leu et al., 2017; 2015).

Properly implementing IT in BPR can have several advantages to the companies. Turnaround time can be reduced by using IT rather than manual approaches which will be more time consuming. Chances of fraud and corruption will lessen and also more quantity of work such as reports can be done in less time. It can produce good quality of work results, services and products, and in a team a quick communication can be formed. Besides this, faster communication can also form with the customer and other stakeholders with the help of IT in BPR (Sudhakar, 2010).

Finance companies can also take advantage by accepting any customer requests at a point which can eliminate customers' multiple calls and reduce the call center volume. IT will help to automatically update each of the customer's accounts as requested by them and eliminate duplicate data entries as well as potential errors (Sungau and Msanjila, 2012). The team and employees from the organization must be properly trained in the applications of IT and any other related technologies, in order to obtain all these benefits from the combination of IT and BPR (Sudhakar, 2010).

However, incorrectly implementing IT in BPR can also lead to some unavoidable disadvantages. The disadvantages occur by having the following criteria or reasons: One of the reasons being the wrong implementation of IT in the rapidly changing business environment which would lead to creating barriers in responding. Successful BPR cannot be achieved by commonly choosing the IT packages just to speed up the process rather than properly reengineering it (Panda, 2013).

Next, failures can sometimes occur when implementing the BPR in a few companies. Although the percentage is very low, it still may be caused by the lack of implementation as the companies spend more time on planning the processes. Incorrectly identified processes for reengineering can also lead to failure in BPR. Moreover, lack of commitment in the top management in the organization can be counted as another reason of unsuccessful IBPR. In addition, some companies might not have an experienced consultant to guide in the right path to the successful steps of BPR in the organization (Sudhakar, 2010). Therefore, these are the reasons due to which failures can occur in BPR in an organization.

9.6 CONCLUSION AND RECOMMENDATION

BPR is a process which brings radical or drastic change to an organization's process. In this report, it has been explained that BPR is a useful method that is capable of making huge changes in an organization's processes in relation to IS, which can be either beneficial or destructive for the organization.

Unfortunately, contradictory to popular belief, BPR does not automatically guarantee that an organization can meet its desired goals. This is due to the fact that in order to succeed, several challenges and issues must be resolved and overcome. There are also critical factors that can determine whether BPR is considered a success or a failure, either during the implementation or during the sustaining phase, in the short run or in the long run. Methodologies and case studies which focus on differences and similarities have also been brought up to view BPR in different perspectives to show that there are different methods for implementing BPR and also the differences in various sectors of organizations which is further elaborated in the literature review section.

BPR may help an organization to focus on better processes such as storing data in a centralized database system in a more practical and efficient way. The organization could also prioritize the most crucial strategic goals to sustain their competitiveness, lower the business threats, be more customer-focused and improve their employees' competency in completing their tasks. However, in order to minimize the resistance to change from the employees and stakeholders, a clear objective, the benefits and expected results should be explained thoroughly, to make them understand the outcomes and offer their dedicated commitment until the BPR project meets the organization's objectives. A proper training should also be provided to these employees as a pilot system to record their difficulties and formulate prototyping before the organization certainly decides to launch the new IS.

KEYWORDS

- **framework**
- **human resource management**
- **information technology**
- **project management**
- **stakeholder**

REFERENCES

Abdolvand, N.; Albadvi, A.; Ferdowsi, Z. Assessing Readiness for Business Process Reengineering. *Bus. Process Manag. J.* **2008,** *14*(4), 497–511. doi:10.1108/14637150810888046

Alghamdi, H. A.; Alfarhan, M. A.; Abdullah, A. L. BPR: Evaluation of Existing Methodologies and Limitations. *Int. J. Computer Trends Technol.* **2014,** *4,* 224–227.

Al-Mashari, M.; Al-Mudimigh, A. ERP Implementation: Lessons from a Case Study. *Inf. Technol. People* **2003**, *16*(1), 21–33. doi:10.1108/09593840310463005

Almunawar, M. N.; Anshari, M.; Susanto, H. Crafting Strategies for Sustainability: How Travel Agents Should React in Facing a Disintermediation. *Operational Research* **2013a,** *13*(3), 317–342.

Almunawar, M. N.; Susanto, H.; Anshari, M. A Cultural Transferability on IT Business Application: iReservation System. *J. Hospitality Tourism Technol.* **2013b,** *4*(2), 155–176.

Almunawar, M. N.; Anshari, M.; Susanto, H.; Chen, C. K. Revealing Customer Behavior on Smartphones. *Int. J. Asian Business Information Management* **2015a,** *6*(2), 33–49.

Almunawar, M. N.; Susanto, H.; Anshari, M. The Impact of Open Source Software on Smartphones Industry. In *Encyclopedia of Information Science and Technology,* 3rd ed.; Khosrow-Pour, M., Ed.; IGI Global, 2015b, pp 5767–5776.

Almunawar, M. N.; Anshari, M.; Susanto, H. Adopting Open Source Software in Smartphone Manufacturers' Open Innovation Strategy. In *Encyclopedia of Information Science and Technology, 4th ed.;* Khosrow-Pour, M., Ed.; IGI Global, 2018b, pp 7369–7381.

Anand, A.; Wamba, S. F.; Gnanzou, D. A Literature Review on Business Process Management, Business Process Reengineering, and Business Process Innovation. Paper presented in The 9th International Workshop on Enterprise & Organizational Modeling and Simulation (EOMAS 2013). http://ssrn.com/abstract=2263123(accessed June 17, 2013).

Cao, G.; Clarke, S.; Lehaney, B. A Critique of Bpr from a Holistic Perspective. *Bus. Process Manag. J.* **2001**, *7*(4), 332–339. http://dx.doi.org/10.1108/EUM0000000005732
Doumeingts, G.; Browne, J. Modelling techniques for business process re-engineering and benchmarking. Paper presented at the IFIO TCS WG5.7 International Workshops on Modelling Techniques for Business Process Re-engineering and Benchmarking, France. https://books.google.com.bn (accessed April 19, 1997).
Habib, M. N. Understanding Critical Success and Failure Factor of Business Process Reengineering. *Int. Rev. Manage. Bus. Res.* **2013**, *2*(1), 1–8. http://www.irmbrjournal.com
Habib, N.; Shah, A. Business Process Reengineering: Literature Review of Approaches and Applications. Paper presented in the Proceedings of 3rd Asia-Pacific Business Research Conference. http://wbiworldconpro.com/uploads/malaysia-conference-2013/management/433-Nauman.pdf (accessed Feb 25, 2013)
Huang, Z.; Palvia, P. ERP implementation issues in advanced and developing countries. *Business Process Management J.* **2001,** *7*(3), 276–284
Jamali, G.; Abbaszadeh, M. A.; Ebrahimi, M.; Maleki., T. Business Process Reengineering Implementation: Developing a Causal Model of Critical Success Factors. *Int. J. e-Educ. e-Bus. e-Manage. e-Learn.* **2011,** *1*(5), 354–355. http://www.ijeeee.org.
Jurisch, M. C.; Ikas, C.; Palka, W.; Wolf, P.; Helmut, K. A Review of Success Factors and Challenges of Public Sector BPR Implementations. *Technische Universitat Munchen* **2012.** DOI:10.1109/HICSS.2012.80.
Jurisch, M. C.; Ikas, C.; Wolf, P.; Krcmar, H. Key Differences of Private and Public Sector Business Process Change. *e-Ser. J.* **2013**, *9*(1), 3–27. http://www.jstor.org/stable/10.2979/eservicej.9.1.3%20.
Kemp, J. Asian Paints Enable Growth Through Improved Planning. *Supply Chain Leader* **2009,** (8), 36–37. http://laurenbossers.writersresidence.com/system/attachments/ files/3629/original/Supply_Chain_Leader_issue_8.pdf.
Leu, F. Y.; Liu, C. Y.; Liu, J. C.; Jiang, F. C.; Susanto, H. S-PMIPv6: An Intra-LMA Model for IPv6 Mobility. *J. Network Computer Applications* **2015,** *58,* 180–191.
Leu, F. Y.; Ko, C. Y.; Lin, Y. C.; Susanto, H.; Yu, H. C. Fall Detection and Motion Classification by Using Decision Tree on Mobile Phone. In *Smart Sensors Networks;* Xhafa, F., Leu, F.-Y., Hung, L.-L., Eds., Academic Press, 2017, pp 205–237.
Liu, J. C.; Leu, F. Y.; Lin, G. L.; Susanto, H. An MFCC-Based Text-Independent Speaker Identification System for Access Control. *Concurrency Computation: Practice Experience* **2018,** *30*(2), e4255.
Marjanovic, O. Supporting the "Soft" Side of Business Process Reengineering. *Bus. Process Manage. J.* **2000,** *6*(1), 43–53. http://dx.doi.org/10.1108/14637150010313339.
Mohapatra, S. *Business Process Reengineering: Automation Decision Points in Process Reengineering;* 2012. Springer Science & Business Media.
Mudiraj, A. R. BPR: The First Step for ERP Implementation. *Int. Res. J. Commer. Bus. Soc. Sci.* **2014,** *2*(13), 1–2. https://www.academia.edu/7166984/BPR_The_first_step_for_ERP_Implementation.
Fui-Hoon Nah, F.; Lee-Shang Lau, J.; Kuang, J. Critical Factors for Successful Implementation of Enterprise Systems. *Business Process Management J.* **2001,** *7*(3), 285–296.
Paint, A. *Annual report 2011–2012.* Asian Paints. 2012. https://www.asianpaints.com/company-info/about-us/corporate-information.aspx.

Project Management Institute. *A Guide to the Project Management of Knowledge: PMBOK Guide*, 5th ed.; Project Management Institute, Inc: Pennsylvania, USA, **2013**.
Sandeep, K. Project report on reengineering building material (Tiles/Paint/Marble) Industry. https://www.scribd.com/doc/37671425/Re-Engineering-Building-Material-TilesPaintMarble-Industry (accessed Nov 18, 2010)
Setegn, D. Assessing the Effect of Business Process Reengineering on Organizational Performance: Economic Development. *Int. Refereed Res. J.* **2013**, *4*(1), 116. www.researchersworld.com/vol4/vol4_issue1_1/Paper_13.pdf
Susanto, H. Managing the Role of IT and IS for Supporting Business Process Reengineering. *J. Systems Information Technol.* **2016a.**
Susanto, H. IT Emerging Technology to Support Organizational Reengineering. **2016b.** https://ssrn.com/abstract=2770318
Susanto, H. Cheminformatics—The Promising Future: Managing Change Of Approach Through ICT Emerging Technology. In *Applied Chemistry and Chemical Engineering, Volume 2: Principles, Methodology, and Evaluation Methods*; Haghi, A. K., Pogliani, L., Balkose, D., Mukbaniani, O. V., Mercader, A. G., Eds.; Apple Academic Press, 2017a, pp 313–332.
Susanto, H. Biochemistry Apps as Enabler of Compound and DNA Computational: Next-Generation Computing Technology. In *Applied Chemistry and Chemical Engineering, Volume 4: Experimental Techniques and Methodical Developments;* Haghi, A. K., Pogliani, L., Castro, E. A., Balköse, D., Mukbaniani, O. V., Chia C. H., Eds.; 2017b, Apple Academic Press, pp 181–202.
Susanto, H. Electronic Health System: Sensors Emerging and Intelligent Technology Approach. In *Smart Sensors Networks;* Xhafa, F., Leu, F.-Y., Hung, L.-L., Eds., Academic Press, 2017c, pp. 189–203.
Susanto, H. Smart Mobile Device Emerging Technologies: an Enabler to Health Monitoring System. In *High-Performance Materials and Engineered Chemistry;* Torrens, F., Balköse, D., Thomas, S., Eds.; 2018, Apple Academic Press. pp 241–264.
Susanto, H.; Almunawar, M. N. Managing Compliance with an Information Security Management Standard. In *Encyclopedia of Information Science and Technology,* 3rd ed.; Khosrow-Pour, M., Ed.; IGI Global, 2015, pp 1452–1463.
Susanto, H.; Almunawar, M. N. Security and Privacy Issues in Cloud-Based E-Government. In *Cloud Computing Technologies for Connected Government;* Mahmood, Z., Ed.; IGI Global, 2016, pp 292–321.
Susanto, H.; Almunawar, M. N. *Information Security Management Systems: A Novel Framework and Software as a Tool for Compliance with Information Security Standard.* CRC Press, 2018.
Susanto, H.; Chen, C. K. Macromolecules Visualization through Bioinformatics: an Emerging Tool of Informatics. *Appl. Phys. Chem. Multidiscip. Approaches* **2018a,** 383.
Susanto, H.; Chen, C. K. Informatics Approach and Its Impact for Bioscience: Making Sense of Innovation. *Appl. Phys. Chem. Multidiscip. Approaches* **2018b,** 407.
Susanto, H.; Almunawar, M. N.; Tuan, Y. C. Information Security Management System Standards: a Comparative Study of the Big Five. *Int. J. Electrical Computer Sci.* **2011,** *11*(5), 23–29.
Susanto, H.; Chen, C. K.; Almunawar, M. N. Revealing Big Data Emerging Technology as Enabler of LMS Technologies Transferability. In *Internet of Things and Big Data*

Analytics Toward Next-Generation Intelligence; Dey, N., Hassanien, A. E., Bhatt, C., Ashour, A. S., Satapathy, S. C., Eds.; Cham: Springer, 2018, pp 123–145.

Trkman, P. The Critical Success Factors of Business Process Management. *Int. J. Inf. Manage.* **2010,** *30*(2), 125–134. http://www.elsevier.com/locate/ijinfomgt.

Tsalgatidou, A. Methodologies for Business Process Modelling and Reengineering, 2014. http://www.appbook.org/read-files/section-2-business-process-reengineering.

Violet, M. S.; Watundu, S. A Framework of Business Process Re-engineering and Organisational Resistance. Paper presented at The ICT Conference, Nigeria, 2011. http://www.orsea.net/

Wanare, R. S.; Mudiraj, A. R. Study on Business Process Reengineering (BPR) and Its Importance in ERP Implementation. *IJRCCT* **2014,** *3*(7), 715–719.

Zigiaris, S. Business Process Re-engineering (BPR) (Report produced for the EC Funded Projects). Lisboa, Portugal: Agency of Innovation. 2000. *Web-6.*

CHAPTER 10

MANAGING ONLINE LEARNING: BIG DATA, SOCIAL NETWORKS, AND CLOUD COMPUTING

CHIN KANG CHEN[1,*], HERU SUSANTO[2], and FANG-YIE LEU[3]

[1]*The Indonesian Institute of Sciences, Indonesia,*
E-mail: chinkang.chen@ubd.edu.bn

[2]*Tunghai University, Taichung, Taiwan,*
E-mail: heru.susanto@lipi.go.id

[3]*University of Brunei, Bandar Seri Begawan, Brunei,*
E-mail: leufy@thu.edu.tw

[*]*Corresponding author. E-mail: chinkang.chen@ubd.edu.bn*

ABSTRACT

In order to understand the depth of managing online learning, this chapter will be discussing the comparison of these three terms: big data, social network, and cloud computing. Meanwhile, it also focuses on the benefits of online learning, reviewing both the positive and negative impacts on users and businesses involved when making online learning as their platform to perform better. The competition between organizations to prove that they are better than the other has become savage, and they need to be on top in order to be noticed by consumers. Hence, the competitions for every business to achieve the lowest cost and at the same time ensuring that their product quality are at the standard required can also be reached easily by making the use of technology. This is when training the staff is very crucial. Therefore, this chapter will be discussing by the use of online learning and how to manage it using big data, social networks, and cloud computing. The authors also decided to do survey based on the use of online learning in Brunei and comparisons to other countries, which will be discussed in Results and Discussion section. As for the discussion, there will also be the impact—the positive and negative effects in using online learning.

10.1 INTRODUCTION

Improvement in technology has highly influenced the society, especially in terms of Internet usage. It has become a trend in the society to use the Internet for the purposes of employment, education, social life, and so forth. With the ever-changing technological advancement and highly increasing usage of the Internet, the amount of data being processed everyday has escalated. Albeit, the insights of "big data" have given the users the efficiency that they need in order to strive for a better competitive edge in this modern era. Furthermore, they have offered the industrialists to recognize each of the problems while resolving them with an immediate feedback and solutions resulting in ceding them with a better competitive advantage to be compared to their rivals.

Profoundly, social network has become a mean for businesses to reach more potential customers while at the same time training their employees in improving their skills. While it helps most organizations to be more productive in a short amount of time, it also has tremendously saved more expenses cost. First, this chapter will briefly introduce to each of the terms and will then further give an analysis of the study.

10.1.1 DEFINITIONS

Big data has become a universal phrase in the world of academia and business organizations. However, the term has ambiguous definition as various people have their own interpretation of what big data actually is. Ward and Barker have studied that generally big data falls under two concepts, which are data storage and data analysis. In addition to the two concepts, Doug Laney has also articulated big data to have three Vs for the characteristics—velocity, volume, and variety—where, these three dimensions refer to the processing of data speed, the amount of data, and the different types of data accordingly.

Social network has been defined as an online service where people can create and maintain interpersonal relationship using Internet. Through social network, its connection is being fully identified and utilized through the web-based groups established for the purpose. The term "social network site" varies with "social networking sites" as networking emphasizes the relationship established between strangers. Its primary use is communicating with people who are part of the extended social network. Thus, to express social network as a critical organizing feature of these sites, it is named as "social network sites."

As technological advancement has slowly raised its standard to reach more potential market, it has used most of the applications as the platform for it to offer services, which is also known as software as a service (SaaS). Nowadays, consumers have become more knowledgeable and their demand has drastically increased. It has not only become a trend but technology emerges to become the necessities, be it for individual or organizations as well. Cloud computing has become the platform as it offers the flexibility of data storage although it is still an evolving paradigm. (Susanto, 2016a; 2016b)

Big data, social network, and cloud computing are often confused to fall under the same category, although these three terms are interrelated to each other. Social network is more toward a platform for people to interact with one another, although the data being posted might be considered to be big data. At the same time, cloud computing may be referred in which users are able to store information.

Online learning should be managed very well and the online learning management manages it. The communication tools used are mostly using technology, such as e-mail, web, and file-sharing application. This is created for the purpose of learning or communication between the management and its teams.

10.1.2 BENEFITS

In big data, in order to extract data from huge amounts of structured and unstructured data, it requires a proper system and an appropriate tool—a precise combination of people, processes, and analytic materials. With these being implemented, some business benefits include (Kuketz, 2012):

1. Big data is timely insights from the vast amounts of data, such as data stored in the company database, from external third-party sources—Internet, social media, and remote sensors

2. Able to identify significant information that can improve decision quality
3. Real-time monitoring and forecasting of events that will give impacts to the business performance or operation
4. Reducing risk by optimizing the complicated decisions of unplanned events more rapidly

As for social networks used for online learning, it gives a lot of criticism of being a distraction to users but it offers plenty of opportunities for learning and interactivity. With the use of technology, users are experiencing, learning, and adapting the world using a relatively new form of communication. It gives benefits in term of (Baker, 2013):

- *Connections*: Social media networks are set as the platforms for the purpose of communal connections. Through social media, users are able to interact and engage with others through the presence of websites.
- *Knowledge*: With social media, users are able to give and receive information at rapid speeds. Their ability to assess, analyze, retain, and share information is accelerating and without any realization, users are not able to realize the skills that they are developing.
- *Social media marketing*: Through the advert and dominance of social media, it has created new variety of marketing, which then involved in professional skills to build it. As users enter the workforce, they also bring their skills to their careers. As social media created young workers into great marketers, businesses have included social media as their marketing strategy.
- *Web engagement*: Students use social media not only for the purpose of social life but also in order to interact with their peers and even their lecturers regarding their studies. As for businesses, "students" are becoming experts in developing sense of Internet presence.

Cloud computing has become popular in education and business settings. With the advance technology, there are high chances in users relying on cloud computing in order to solve the challenges faced by small business, such as checking bank balance or using e-mails to send or receive messages (Liu et al., 2018; Leu et al., 2017; 2015). Its benefits are (Sales Force UK, 2015):

- *Work from anywhere:* Using cloud computing, users are able to do their works or studies from anywhere with the use of Internet connection. Cloud services offer mobile apps, which users are mainly have access to phone and is not restricted to use the devices. As a result, businesses can offer flexible working perks to their employees (Susanto, 2018).
- *Security:* A greater loss is when the users lose their devices, which leads to a greater loss of sensitive or confidential data. With the use of cloud computing, it gives security to your data as it is stored in the cloud, and users are able to access them despite the loss of their devices, and so forth. Users are also able to remove their data from their lost devices so that it will not get into the wrong hands (Susanto, 2018; Susanto, 2011).
- *Disaster recovery:* It is ideal for business to invest in disaster recovery, but as for small business with monetary issues, cloud is able to help these businesses to use the service. This also can help the businesses to avoid large investment and the use of third-party expert to deal with this situation (Susanto and Almunawar, 2016).
- *Automatic software updates:* Cloud computing can be off premise, where it can be out of sight. Suppliers deal with the matters and release regular software updates—including security updates. This gives the users less problem to maintain the system and able to focus on other matters such as expanding the business (Susanto, 2016a; 2016b).

10.1.3 CRITERIA

Since online learning has such great impact on students, there are several criteria that they need to meet in order to make the learning more effective. Humans have the need to feel accepted, and therefore, their participation level may vary according to their environment. Cultural, physiological, social, and physical factors are some of the factors that are conducive to the students' learning activities. Some of the characteristics that contribute to making better learning environments are as follows:

- *Safe:* Educators ought to provide safe surroundings for the students to feel at ease when having a group discussion with the peers and lecturers. Rules of online correspondence, for example, netiquette, can set up a safe online environment that is free from individual assault while having abundant in important and vital information. The students will also be given the freedom to speak up while acknowledging their other commitments apart from just learning obligation.

- *Supportive:* Building a supportive environment for the online students is also a vital part in ensuring that the students feel more comfortable in sharing ideas on the web. Students ought to be urged in making inquiries provided that if they have any problem on certain topic, they will be able to start an online discussion with the peers in order to help one another through online talk and virtual understudy lounge. This can have a major impact in helping understudies remain focused while succeeding in web learning.
- *Interactive:* Online understudies for the most part do not have the chance to speak with their educator and companions up close and personal. Therefore, to make the online learning more successful, it is essential to make Internet learning more interactive so that the learners will not feel disengage with one another. The potential outcomes for expanded cooperation in the online classroom will keep on growing with synchronous devices. Three sorts of connections are vital for web learning: communication among students and materials, students and students, and students and educator.
- *Flexible and engaging:* Online learning enables students to learn at any place according to their own pace. This gives the students a feeling of control over their own learning. Furthermore, it is crucial to permit understudies to seek after their subjects of interest or customized ventures. It is a decent approach to propel online learning. At the point when understudies impart their individualized work to peers, every understudy is presented to a more extensive range of subjects and can gain from kindred understudies. Furthermore, when the other students share their piece of work with the entire class through the web, their peers might learn more in a broader spectrum from them.

All in all, to ensure that natural components are embedded in every phase of the instructional outline, lecturers or educators must be able to provide the students with four criteria of environment that will make online learning more effective and improving number of performances.

10.2 METHODOLOGY

This section will be examining on how the information regarding managing online learning by using big data, social network, and cloud computing in this dissertation is collected. Both primary and secondary sources are utilized to make the data more reliable and accurate. Authors will be defining the methods enforced while this study of this case is being conducted.

The respondent were asked the basic questions such as their awareness of big data, social networks, and cloud computing. Furthermore, the respondents were also asked the purpose of the user for using Internet.

10.2.1 PRIMARY SOURCE

The authors of this chapter have taken the quantitative approach in collecting the information, whereas, it is a formal research where data is collected directly from a specific sample group of approximately 50 people ranging from 18 to 45 years of age. The authors have chosen to conduct survey because it is easy to gather data in a limited amount of time while at the same time gathering the information needed to be used in this chapter. Moreover, authors will be able to group the individuals while concluding the overall outcome in percentage form.

This can provide comprehensive portrayal of the study in which managing online learning through big data, social network, and cloud computing has been made easier for the society to thrive better.

10.2.2 SECONDARY SOURCE

Secondary information is the data that have been gathered by and promptly accessible from different sources. Such information is less costly and can be rapidly reachable than the essential information. Furthermore, data can be obtained when primary information cannot be gathered by any means or simply to give more reliability to the dissertation.

Authors summarize the global descriptions collected from the scholar published articles and are to be used in supporting finding of the case.

10.3 LITERATURE REVIEWS

It is discovered by Akyildiz and Argan in one of their studies that social network has been playing the role of preexisting social relations, social interaction, business strategies, and learning and teaching (Akyildiz and Argan, 2012). It provides us the overview that social network has taken over the traditional way of human interaction as people are more relying on the use of technological portable devices in order to communicate.

With the growth of information technology (IT), investigation that has been conducted by Docebo, shows that, there were comparisons of online learning market in regions. Focusing on Asian market, Asian has the world's fastest regional growth rate of 17.3%. The sales revenue in online learning reached $5.2 billion in 2011 and is expected to double by the end of this year. With the increased penetration of Internet, the low-cost and rising demand helps the market to develop strongly in the future. It is no longer a trend but a necessity in order to be able to gain the competitive advantage in today's world (Docebo, 2014).

In cloud computing, the online learning system includes software components, such as client application, an application server, database server, and necessary hardware components. Client application can be accessed using mobile device or computer in which a simple web browser or dedicated application is installed.

10.3.1 EDUCATION

A study by Drennan et al. (2010) has shown that students are more interested in learning by using IT as their platform to learn as they have an easy access through their own portable technology devices at any time. Furthermore, it has also broadened their mind as they are practicing the nontraditional way of learning. The students are being more exposed on how to properly manage their information technological skill, which will be beneficial for them in near future. Moreover, younger generation is open-minded and is not reluctant to change when it comes to using the new technology. This leads them to discover an increase in not only satisfaction of the students increase but also the performance of the students over the year.

However, Wright (2014) stated that even with the use of online learning, students have not established any vital boundaries in the early online learning process, where students would not be able to differentiate how to engage in a mystery bland text communication, how much information they are exposing (confidential information on each individual) and being shared on the net. Thus, he argued that students should be educating themselves on how to safely share their confidential information so that it will not be stolen or shared without their consent.

In an online report card, composed by Allen and Seaman (2016) stated that institutions in the United States identified online education as a critical point in their long-term strategy. In the survey, there was a decline pattern from 2012 to 2015 among them who were neutral to this statement. Between these years, the pattern of people agreeing to this statement was fluctuating and as for the people who disagree to it, there was a decrease pattern, but toward 2015, there was an increase from 8.6 to 13.7%. Despite the largest drop of percentage in online learning as a critical point in their long-term strategy, it does not mean that the institutions will abort the online courses and programs (Allen and Seaman, 2016).

According to Bosch (2009), social network make it easier for the students to actually access the materials for their subjects easier than the traditional form of learning. Apart from that since the new generations are more exposed to go green initiative, they are more than willing to use social network as platform for them to learn. Social network also help students who are shy to interact in class to be more active in participating to virtual discussion with the other students (Bosch, 2009).

Despite the easy access of using social networks for education, the lecturers from the University of National and World Economy (UNWE) found that the approach of using social networks as a tool of education is not applicable as they see social networks are basically children's toy, where people use social network for the purpose of interaction and social life.

Furthermore, Garrison and Innes (2005) stated that in using online learning, the relationship of interaction and the presence are still lacking. They stated that even if there are interactions, it is not guaranteed that the students are cognitively engaged in the education in a meaningful manner.

While finding other ways to improve the agility of educational institutes, Mircea and Andreescu(2011) discovered that one alternative way to improve the growing needs of providing necessary IT is by using cloud computing. This strategy does not only correspond to the problems that each universities are facing but are also able to restructure the university's architecture and improving the knowledge field in this area.

According to Pocatilu et al. (2009) with the constantly changing needs of education involving technology and development in terms of cloud computing, e-learning system hasbeen the most effective ways for education, especially for the students. It reduces the cost of using resources as it is convenient for the students to access it anywhere at any time and infrastructure of the university in terms of classroom-based learning.

Archee and Gurney (2012) supported e-learning has become a standard way in supporting the students in the higher education. Despite the lack of face-to-face interaction, it somehow suits students who prefer to work and study.

A survey was conducted in UNWE where the questions were in terms of general usage of IT as a learning process, the respondents' attitude toward the use of social networks for an educational purpose and their expectation from the implementation of online learning. From the survey, 85% of the participants used social networks as a mean to connect with their peers, while 26% use online communication with their professors.

Ivanova et al. (2015) conducted a study on the use of cloud technologies by the students. A survey was conducted in the Bulgarian universities. About 50% of the respondents use cloud technologies to access their resources, while 40% of the respondents use cloud technologies for the purpose of education.

10.3.2 ORGANIZATION

Waters et al. (2009) found that ever since technology advanced and social networks are widely used by society, organization found that it is the best strategy to use social networks such as MySpace and Facebook to be implemented into their public relations programming. By this, the organization not only can help to launch their products but also to advance in terms of their organization's mission and programs.

Bughin et al. (2010) mentioned that some businesses, such as Global Energy Services company, have difficulties for geographic unit and business unit and it prevents their managers to solve the client's problem. Thus, with the use of social-network analysis and with the easy access of information flow and knowledge resources among the staff, it makes it easier for the managers to execute the information.

Govindasamy (2001) stated in corporate training institutes that the use of e-learning is efficient in solving authentic learning and performance problems. It is crucial for the e-learning to be implemented successfully as it will bring the return of investment to the organization. Thus, organization is needed to be aware and considerate in underlying pedagogy, or how to use online learning.

Nonetheless, Archee and Gurney (2012) specified that it can be a major concern using e-learning to transfer media files as network can have problem such as network down or server down.

In order for organization to be able to maintain its continuous education using online learning, the organization needs to be able to observe their expenses in terms of time and difficulties. The training activities need to be monitored and managed using stable and dependable tracking system that can be stored, analyzed, and consulted. This training management known as learning management system is the key to the professional development plan as well as the organization's human resources strategy (Docebo, 2014).

10.4 RESULTS AND DISCUSSION

This section will discuss about the survey conducted on 55 respondents on managing online learning through big data, social network, and cloud computing. The authors managed to collect data from 39 females, while the others are males. As this survey is being conducted online, it also shows that females are not reluctant to use this mean as a platform for answering survey which also relates to the topic of the case study. Under the discussion section, there will also be given the impact—negative and positive effects, using online learning in social networks and cloud computing.

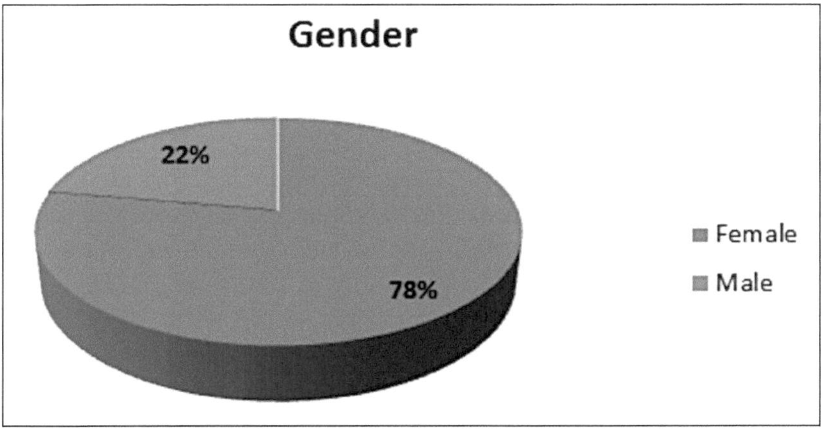

FIGURE 10.1 (See color insert.)

The majority of people answering this survey are from the age of 18 years to 29 years, that is, 90.7%. This is because younger people tend to have their phone in closer proximity all the time, hence, the result.

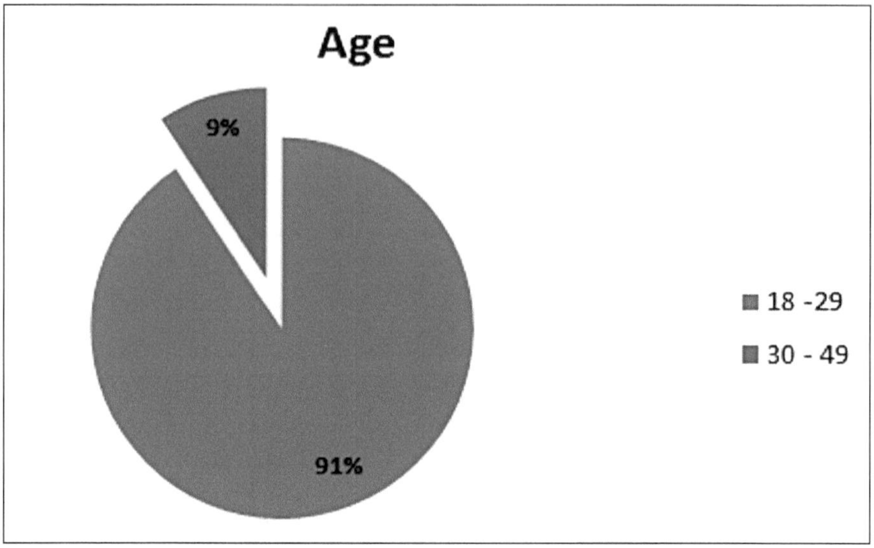

FIGURE 10.2 (See color insert.)

Occupations for the respondents are also considered in order to investigate whether people who are working or studying are more aware of the topic being studied. The authors have divided the working section into three parts, which are government sector, private sector, and self-employed. About 72.2% of the students answered the survey, while 14.8% from the government sector, 11.1% from the private sector, and lastly, 1.9% of the total respondents are self-employed.

From the survey, about 31.5% respondents are familiar with the terms big data, social network, and cloud computing, while 63% of them are familiar with one or two terms given, and the others are not aware with the terms at all. Despite this, respondents are still unable to differentiate these terms and are still unsure about it. Such that 90.7% of the respondents said that they have used one of these terms, while 9.3% have not. It cannot be concluded that Bruneian are illiterate with these terms as the respondents are mostly from the age of 18 to 29 years.

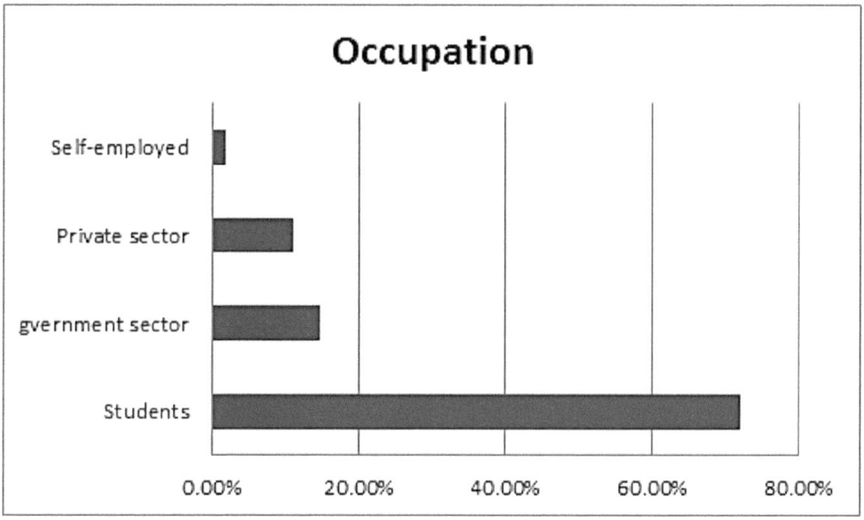

FIGURE 10.3 (See color insert.)

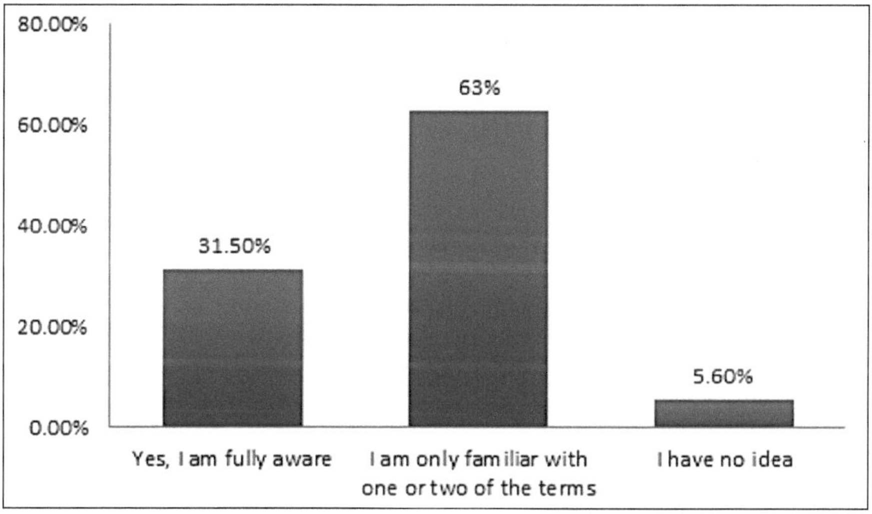

FIGURE 10.4 (See color insert.)

Among the big data, social networks, and cloud computing, 60.4% respondents are using social networks for online learning purposes. It can be stated that since social networks such as Facebook, Twitter, WhatsApp, and so forth, are much easier and accessible to use, especially using tablets, laptops, and mobile phones (Susanto et al., 2016). Moreover, the additional features inside the application make the social network more fun to use, and hence, attracting more people to use it (Almunawar et al., 2018a; 2015a; 2015b).

For example, online learning using Facebook—it is not only for socialization but Facebook has been updated as well and has new features as months passed by. Developing these features and performance enable users to not only interact and build new friendship with strangers, but theyalso make it easier for lecturers, students, or business partners to interact, send, receive, or transfers information to each other.

Using Facebook, users will be able to upload articles that would support their personal development such as careers tips, self-assessment tools, and also industry news. Furthermore, additional resources can be obtained such as e-books, videos, presentations, and so forth, to assist students in completing their programs and getting them interested in other programs within the brand.

Online learning providers are also able to interact with their users In such a way that the providers will assist the users in motivating to finish their courses or solving any issues in user queries by providing additional guides and instructions. With the easy interaction, users-to-users communication is built in this c-learning space and engage in online community. Users can share their experiences, knowledge, and stories and limit any provoking participation.

Growth of technology has been rapidly increasing globally, including Brunei itself, it can be concluded that Bruneian has easy access to the Internet, thus able to use Internet for the purpose of education, socialization, entertainment, and business. Mobile technology, as one of the technologies used in Brunei, is playing a major role in this entire digital experience. Despite the mobile ecosystem in Brunei such as the devices, carriers, the apps, and so forth, has grown in a slow pace, but the use of Internet, especially online learning, is still utilized around the country and yet needed to be educated with the use of online learning.

In organization, the use of online learning and its demand for customization in the online learning content and technology has drastically increased to facilitate the talent management. Thus, in the future, the probability for the online learning growth is expected to increase the market (Susanto et al., 2018).

In cloud computing, users use mobile devices for learning more than desktop computer. Although there are limitations in mobile devices in term of current hardware and software, mobile devices support multimedia-based applications. As for the desktop application, when compared with mobile devices, the serious limitations are the processing power and its memory constraints. Thus, online learning using mobile devices are developed faster (Susanto, 2017a; 2017b; 2017c; Susanto and Chen, 2017).

To understand further more on big data, social networks, and cloud computing, listed below are the impacts—negative and positive use of online learning with additional discussion on these points. In this context, students are used not specifically for them to attending school. It is used as general including in training programs in organization also.

Big data as a brief is the data from traditional and digital sources inside and outside of the company that represents the source for analysis, comprising unstructured data, where information is not organized by traditional database and multi-structured data, where there are varieties of formats and types of data that can be derived from interaction between people and machines. Typically, big data is collected at specific intervals but it can be collected and analyzed constantly. Thus, there are several advantages and disadvantages users should know on big data:

- It is secure, in terms of its infrastructures being built by big data where it is hosted and technology partners can protect the organization revenue. The downside to this is security breach. Thus, once the security is broken down, all the organization confidential information will get to wrong hand.
- Even though big data is enormous, users tend to misunderstand this term and somehow frightened with this word. With the enormous data, it means that the platform carry unlimited information, making it easier for users to search and use their data or information. The drawback of this is that the conflict for the users to fetch the relevant information—numerous data or information can lead to confusion to the users.
- Moreover, in big data, every minute and every second, there will always be new information published on the platform. In other words, big data has a speedy updates. The drawback of this is that the updates can be a mismatch to the real figures. By mean with the constantly updated data, it can be an error in the information.

- As for organization, big data has an advantage with respect to the competitive point of view. Big data is a real-time analysis where it allows businesses to develop more of its effective strategies toward competition in less time. It also offers deep insight into consumer trends and sales. The downside to this is that this real-time big data demands the ability to conduct sophisticated analyses. Such organization who fails to do so will implement incorrect strategies of the whole organization (Almunawar et al., 2015; 2015a; 2015b).

As mentioned above, social network has become one of the trends for users for the purpose of online learning. This includes the use of online learning in Brunei itself. With the spreading use of technology using mobile devices, users, especially, students agree to this condition as it is accessible. Users needed to be aware of the advantages and disadvantages of using social networks as their online learning equipment:

- It is known that social network has increased the student collaboration, especially, becoming friends closer with their classmate using this medium. With this, students have high retention rates as they are more connected with their institutions.
- Using social networks, faculty from the institutions can also able to interact with students such as the students who are shy or unable to communicate fluently in their classroom, also, faculty will be able to listen to any queries from the students. For example, faculties in Universiti Brunei Darussalam has created Facebook page in order to make it easier for the students to get information precisely and make it visible to other students for any issues brought up by a student. Nonetheless, this does not only imply to students but also to the organization, where they use social networks as their platform to discuss their projects and so forth.
- Furthermore, social networks provide massive resources, such as video, websites, tutorials, or training programs that can be shared easily in the entire classroom. This can reduce the errors that are made by students such as error in copying link or mistyping. With the use of social networks, information given is easily accessed just by one click.

The disadvantages are:

- With the use of social networks, students can easily get distracted. Instead of educational purpose, students would be easily preoccupied doing their personal work and not paying attention during their classes. Neither lecturers nor the faculties are able to monitor these activities.
- Even though social networks have advantages to the shy students in the classroom, it gives a drawback to them. Such that students will have problems to communicate in face-to-face situation. This has somehow become a problem to the society and limits the student's education.
- Security issues can also be encountered by the users, such as the possibility of hackers or unauthorized users that would result in identity theft, loss of data and its productivity, and so forth.

Recent evolution in Internet has moved the web pages to environment that allowed user to run software applications. Thus, cloud computing has increased its growth in the IT sector area, where the applications are run in variety of hardware devices, while data storage remains inside the cloud. Here are the advantages and disadvantages of cloud computing in online learning:

- In order to deploy the software and hardware, some companies and education providers have difficulty using these services in term of cost. Thus, cloud technology provides a rental processing capacity that they need not use the infrastructure fully to run the high-quality online learning. Such providers, for examples, are Blackboard and Moodle, which have this version in their packages. This also applies to large organizations, such that they only need fewer servers and data to store in the cloud.
- The use cloud computing also lowers the maintenance issue. It reduces the hardware and software maintenance for the companies because fewer serves are needed and used by the companies. Moreover, the software is stored in the cloud, thus the IT staffs have no software to maintain or upgrade it.
- Cloud computing also affects the performance of the computer. In this case, it boosts up the performance of the computer. It is

necessary to consider that applications are accessed from the cloud, thus, there are less programs and process loaded in the computer memory, which leads to faster start-up and run smoothly.

- Cloud computing also provides unlimited storage capacity. With the use of computer or server on a network, the user is given limited storage to store, whereas, cloud computing has offered unlimited storage capacity.
- It is easier to share information in group collaboration. With cloud computing and easy access to Internet connection, users can collaborate in real time. It enables multiple users to cooperate on the projects or documents at the same time. Even though this does not limit the places, just as mentioned above, it can be accessed in different places or countries easily.

Despite these advantages, there are also several drawbacks to it, for example:

- Even though it is easy to access cloud computing using devices, the connection or network can be an issue. Without connection to access the applications or the documents, users are not able to access it, even their own files. Thus, it makes cloud computing complicated to be accessed in places where there is no dependable connection or network.
- Furthermore, this also includes poor connection that will limit this accessibility. For example, in remote regions, which offer dial-up services, the application or documents are impossible to run, especially on intense duty. As for web-based or large files, it requires a heavy bandwidth to be downloaded. Thus, cloud computing needs to have a stable and fast connection of the Internet.
- Security can also be problem to users. As we know that, through cloud computing, such that information or data are kept in cloud server. It may lead to unauthorized user to access such data through internet. Here, the standard of information security may apply to prevent unauthorized user to take the data from cloud.
- The cost can also be one of the drawbacks, such that over time of data subscriptions, it would increase and it would become more expensive than buying the hardware and software.

Quality of the services from the providers. Even though the provider set the cost of the service less expensive, users need to be aware of the quality. There is a high chance that the services can crash any time and this would leadto inaccessibility to the cloud without any warning and users will not be able to control the condition.

10.5 CONCLUSION AND RECOMMENDATIONS

In conclusion, with the worldwide expand in technology, it indicates the use of Internet, users—students, lecturers, businessman, and so forth—prefer and found the use of learning online be more realistic and not time consuming as Internet is widely available and can be accessed using phone devices, tablets, laptops, and other electronic devices. Despite the drawbacks mentioned in this chapter, it is possible for users in this era to be illiterate in using Internet for the purpose to search, entertainment, or work related. The only issues would be to overcome every situation such as the poor network in rural areas, the cost of the data storage, the security breach, and so forth. Every issue can be handled and solved.

For the next paragraph, here are recommendations for organization, individual, or institution managing the online learning.

As for the students, lecturers or faculties have an important role in the student's learning lifestyle, especially in the use of online learning. Here are few suggestions for lecturers, faculties, or the institution to develop and support student's online learning:

The institutions should outline the students' expectations and by this, when their expectation and roles being articulated, then students should have their own responsibilities.

Next are the queries from students. This is where institutions should be willing to help the students when they have queries. With the applications or software for online learning, some students might find it new and unable to use it at first, or perhaps the students unable to use their account or unable to connect to their courses.

Moreover, institutions should direct the students in order to develop the student's skill. This is in terms of the resources provided by the institution. The resources should be complete and excellent and should be informed to the students earlier.

Lastly, the communication and the information flow needs to be well managed. The communication should be transparent so that the information flow is supported in the Internet environment. The level of awareness concerning project developments should be maintained among the stakeholders. Online learning project members need to do demonstration more by sharing information as much as possible so that the stakeholders make relevant decisions.

KEYWORDS

- **cloud computing**
- **communication**
- **organization**
- **social media**
- **social networking**

REFERENCES

Akyildiz, M.;Argan, M. Using Online Social Networking: Students' Purposes of Facebook Usage at the University of Turkey. *J. Technol. Research* **2012,** *3,* 1.

Allen, I. E.; Seaman, J. *Online Report Card: Tracking Online Education in United States,* February 9, 2016.

Almunawar, M. N.;Anshari, M.;Susanto, H.; Chen, C. K. Revealing Customer Behavior on Smartphones. *Int. J. Asian Business Information Management* **2015,** *6*(2), 33–49.

Almunawar, M. N.; Susanto, H.; Anshari, M. The Impact of Open Source Software on Smartphones Industry. In *Encyclopedia of Information Science and Technology,* 3rd ed.; Khosrow-Pour, M., Ed.; IGI Global, 2015, pp 5767–5776.

Almunawar, M. N.; Anshari, M.; Susanto, H. Adopting Open Source Software in Smartphone Manufacturers' Open Innovation Strategy. In *Encyclopedia of Information Science and Technology, 4th ed.;* Khosrow-Pour, M., Ed.; IGI Global, 2018a, pp 7369–7381.

Archee, R.; Gurney, M. Integrating Culture with e-Learning Management System Design. In *Virtual Learning Environments: Concepts, Methodologies, Tools and Applications;* Information Resources Management Association, Eds.; IGI Global, pp 1389–1405).

Baker, J. How Students Benefits From Using Social Media. February 12, 2013. Edudemic: Connecting Education and Technology. http://www.edudemic.com/how-students-benefit-from-using-social-media/ (accessed Sep 16, 2016)

Bosch, E. T. Using Online Social Networking for Teaching and Learning: Facebook Use at the University of Cape Town. *Communicatio* **2009.** http://dx.doi.org/10.1080/02500160903250648

Bughin, J.;Chui, M.; Manyika, J. Clouds, Big Data and Smart Assets: Ten Tech-Enabled Business Trends to Watch. *Mckinsey Quarterly***2010**, *56,* 75–86.

Docebo. E-Learning Market Trends and Forecast 2014–2016 Report, March 2014. 16.

Drennan, J.; Kennedy, J.; Pisarski, A. Factors Affecting Student Attitudes Toward Flexible Online Learning in Management Education. *J. Educ. Res.* **2010,** *98,* 331–338.

Garrison, D. R.; Cleveland-Innes, M. Facilitating Cognitive Presence in Online Learning: Interaction is Not Enough. *Am. J. Distance Edu.* **2005,** *19*(3), 133–148.

Govindasamy, T. *Successful Implementation of E-Learning: Pedagogical Considerations. Internet High. Educ.* **2001,** *4,* 287–299.

Ivanova, A.;Stoykova, V.;Ivanova, G. In *Social Networking in Higher Education–Good Practices and a Case Study from Bulgarian Universities*, Proceedings of e-Learning'15 International Conference, Berlin, Germany, 2015.

Kuketz, D. The 7 Biggest Business Benefits from Big Data. October 23, 2012. Utopia: Perfect Data, Perfectly Possible. http://www.utopiainc.com/insights/blog/381–7-biggest-business-benefits-from-big-data (accessed Sep 16, 2016).

Leu, F. Y.; Liu, C. Y.; Liu, J. C.; Jiang, F. C.; Susanto, H. S-PMIPv6: An Intra-LMA Model for IPv6 Mobility. *J. Network Computer Applications* **2015,** *58,* 180–191.

Leu, F. Y.; Ko, C. Y.; Lin, Y. C.; Susanto, H.; Yu, H. C. Fall Detection and Motion Classification by Using Decision Tree on Mobile Phone. In *Smart Sensors Networks;*Xhafa, F., Leu, F.-Y., Hung, L.-L., Eds., Academic Press, 2017, pp 205–237.

Liu, J. C.; Leu, F. Y.; Lin, G. L.; Susanto, H. An MFCC-Based Text-Independent Speaker Identification System for Access Control. *Concurrency Computation: Practice Experience* **2018,** *30*(2), e4255.

Mircea, M.; Andreescu, A. I. Using Cloud Computing in Higher Education: a Strategy to Improve Agility in the Current Financial Crisis. *Communications of the IBIMA.* **2011.** DOI: 10.5171/2011.875547

Pocatilu, P.;Alecu, F.;Vetrici, M. *Using Cloud Computing for e-Learning Systems*, Proceedings of the 8th WSEAS International Conference on Data Networks, Communications, Computers World Scientific and Engineering Academy and Society (WSEAS), November 2009, pp 54–59.

Sales Force UK. Why Move To The Cloud: 10 Benefits of Cloud Computing. November 17, 2015. Salesforce UK & Ireland Blog. https://www.salesforce.com/uk/blog/2015/11/why-move-to-the-cloud-10-benefits-of-cloud-computing.html (accessed Sep 16, 2016).

Susanto, H. IT Emerging Technology to Support Organizational Reengineering. **2016a.** https://ssrn.com/abstract=2770318

Susanto, H. Managing the Role of IT and IS for Supporting Business Process Reengineering. *J. Systems Information Technol.* **2016b.**

Susanto, H. Cheminformatics—The Promising Future: Managing Change of Approach Through ICT Emerging Technology. In *Applied Chemistry and Chemical Engineering, Volume 2: Principles, Methodology, and Evaluation Methods*; Haghi, A. K., Pogliani, L., Balkose, D., Mukbaniani, O. V., Mercader, A. G., Eds.; Apple Academic Press, 2017a, pp 313–332.

Susanto, H. Biochemistry Apps as Enabler of Compound and DNA Computational: Next-Generation Computing Technology. In *Applied Chemistry and Chemical Engineering, Volume 4: Experimental Techniques and Methodical Developments;*Haghi, A. K., Pogliani, L., Castro, E. A., Balköse, D., Mukbaniani, O. V., Chia C. H., Eds.;2017b, Apple Academic Press, pp 181–202.

Susanto, H. Electronic Health System: Sensors Emerging and Intelligent Technology Approach. In *Smart Sensors Networks;* Xhafa, F., Leu, F.-Y., Hung, L.-L., Eds., Academic Press, 2017c, pp. 189–203.

Susanto, H. Smart Mobile Device Emerging Technologies: an Enabler to Health Monitoring System. In *High-Performance Materials and Engineered Chemistry;* Torrens, F., Balköse, D., Thomas, S., Eds.; Apple Academic Press, 2018, pp 241–264.Apple Academic Press.

Susanto, H.; Almunawar, M. N. Security and Privacy Issues in Cloud-Based E-Government. In *Cloud Computing Technologies for Connected Government;* Mahmood, Z., Ed.; IGI Global, 2016, pp 292–321.

Susanto, H.; Chen, C. K. Information and Communication Emerging Technology: Making Sense of Healthcare Innovation. In *Internet of Things and Big Data Technologies for Next Generation Healthcare;* Bhatt, C., Dey, N., Ashour, A. S., Eds.; Cham: Springer, 2017, pp 229–250.

Susanto, H.; Almunawar, M. N.; Leu, F. Y.; Chen, C. K. Android vsiOS or Others? SMD-OS Security Issues: Generation Y Perception. *Int. J. Technol. Diffusion* **2016,** *7*(2), 1–18.

Susanto, H.; Chen, C. K.; Almunawar, M. N. Revealing Big Data Emerging Technology as Enabler of LMS Technologies Transferability. In *Internet of Things and Big Data Analytics Toward Next-Generation Intelligence;* Dey, N., Hassanien, A. E., Bhatt, C., Ashour, A. S., Satapathy, S. C., Eds.; Cham: Springer, 2018, pp 123–145.

Waters, D. R.; Burnett, E.; Lamm, A.; Lucas, J. (2009). Engaging StokeholdersThrough Social Networking: How Nonprofit organizations are Using Facebook. *Public Relat. Rev.* **2009,** *35,* 102–106.

Wright, D. R. *Student-Teacher Interaction in Online Learning Environment.* IGI Global: Texas, 2014.

CHAPTER 11

GREEN INFORMATION AND COMMUNICATION TECHNOLOGY AWARENESS AS A CORPORATE STRATEGY FOR INFORMATION SYSTEMS

HERU SUSANTO[1,2], FANG-YIE LEU[2], and CHIN KANG CHEN[3]

[1]*The Indonesian Institute of Sciences, Indonesia,*
E-mail: chinkang.chen@ubd.edu.bn

[2]*Tunghai University, Taichung, Taiwan,*
E-mail: heru.susanto@lipi.go.id

[3]*University of Brunei, Bandar Seri Begawan, Brunei,*
E-mail: leufy@thu.edu.tw

ABSTRACT

Throughout the past decades, the utilization of information and communication technology (ICT) has been escalating by making people's lives and work much more efficient or effective, however, it has also been greatly exploited. ICT has been continuously advancing but problems and questions of these technologies harming our environment negatively are also rising. ICT refers to any technological devices that allow the access to information through telecommunication, while information system (IS) can go by the description as a computer system which has elements that collects, creates, stores, processes, and disseminates with a feedback mechanism (Leu et al., 2015; 2017) . IS is widely used in government organizations, private organizations, or businesses. It can perform business processes such as data storage, corporate website, corporate e-mail, customer transactions processing, finance, accounting, payments processing, payroll processing, sales, marketing, and office tools such as Words and Excel. Subsequently, green information and communication technology (green ICT) designates to minimize the harm that ICT is causing to the environment by making ICT sustainable, waste less energy, disposing it appropriately, using it efficiently, and making it last longer. It is important for users to be aware of green ICT to make the world a sustainable place as global warming or the climate change is rising. It is undeniable that the world is becoming too reliant on technology where most fields are using it (Susanto, 2018; Susanto et al., 2018; Almunawar, 2015; 2018b). Efficiency and effectiveness are achieved through the use of ISs when compared to the past. For example, in a supermarket business equipped with a cash register system with barcode scanning function technology, things would be easy when there is no price tag as cashier can simply scan the barcode to obtain the price, while comparing to the old method, where staff needs to find out the price through the higher management or a recorded system, be it a separate computer or through a book where price is kept. This traditional process causes inconvenience to the customers and also increases the inefficiency in the workplace (Susanto, 2016a; 2016b).

11.1 INTRODUCTION

Despite ICT bringing benefits to its users, it is also a major contributor to the contemporary environmental issues and also has several technological issues. The level of energy consumption from the technology and the rising carbon emission from it is causing a great harm to our environment. In a PowerPoint presentation by Professor G. N. Pandey from Indian Institute of Information Technology, standardization of green ICT with Green communication, he presented a present statistics that showed the current power consumption in ICT is 2%. He also stated that green ICT falls into one of the two broad classes: First, those who reduce greenhousegas emissions with the intention to deal with global warming. Second, establishing an economy with a sustainable growth with technologies associated including recycling, reducing resource and some aspects through biosciences (Susantoand Chen, 2018a; 2018b; Susanto, 2017a; 2017b; 2017c). Green ICT is not a product but it requires the world to work together to save the planet by reducing the use of paper and recycling energy. Moreover, there is no certain way to dispose unwanted or damaged technology units to a point where it is impossible to repair it anymore because it does not decompose at all. Especially, in Brunei Darussalam, there is no certain way to dispose it. It will only damage our environment further when the hazardous chemical or the metal in the technology is not handled well. Griffith (2016) stated that when technology is unusable and cannot be recycled anymore to a point where even charity does not accept it, it is called an electronic waste (e-waste), and it has it could possibly be dangerous to the environment. He further mentioned that electronics are filled with carcinogenic chemicals and heavy metals or toxic metals that are capable to be used but not when they are recycled incorrectly and placed in a landfill. He also stated that thousands of tons e-waste are sent to overseas annually to countries such as India and China, where it gets disposed or possibly burnt, which contributes lead and mercury into the air. It is already dangerous as the environment is deteriorating but some human actions are still contributing profusely due to people's ignorance on the changing climate or global warming. Ahmed et al. (2016) suggested reducing ICT's own energy consumption through the adoption of the SMART concept to measure the energy consumption:

1. Standardization of energy measurement method where the first requirement toward the energy saving is to know how much it had been used. The measuring technique has to be uniform and effective.
2. Monitoring of the energy used. He stated that to obtain a full picture of the energy consumption, we can monitor through the sensors and meter in the segment of technology.
3. Accounting for energy at each step because it is not used only at the end equipment or the server but it should be reported.
4. Researching and rethinking the revolution to lessen the ICT's emissions from the services or the devices where the constant change has caused a small and light part in the reduction of energy consumption.
5. Transformation of ICT sector into an epitome of low carbon technology where this sector needs to have the capability to perform highly efficient system by going into the path of green technology and lead others to follow the path of green technology.

Furthermore, the efficiency and effectiveness ofan organization can be affected when the technology breakdown or hacking causes an organization to lose all its information. The real question is what will happen to organizations that are too dependent on ICT? When the technology used is experiencing an error, people would usually go back to pen and paper and do things manually as a contingency plan. On the other hand, the cost of maintaining the system could be handful as such technologies are not cheap to maintain, especially when the problems are difficult to diagnose. Breach of authorization in an organization canalso be a problem when unauthorized personnel access information that is authorized to certain people even when there is a system that only allows authorized personnel to access. These are the few examples of disadvantages that an organization can experience where it can also slow down the efficiency and effectiveness of using ICT.

In this research, we have obtained information from two government organizations (organizations A and B) and four private-level organizations in Brunei Darussalam (organizations C, D, E, and F) with forty participants through a survey questionnaire provided. The predetermined survey questionnaires were divided into four sections consisting of the IS or ICT awareness, green ICT, attitudes toward green ICT, and the business process reengineering (BPR) awareness for business sustainability. The organizations and participants will be denoted as organizations A, B, C, D, E, and F in the report where the organized responds from the organization will be represented in word form and graph figure. About 40 participants' position title that answered the survey differs from organization to organization where it ranged from the top-ranking management to the lowest ranking employee.

The purpose of this report is to give a detailed outline on our findings and brief discussions from what we have obtained through six local organizations on green ICT through IS in the government- and business-level sectors. The research survey was aimed to demonstrate the influence of green ICT in BPR and the conveniences that the organizations had attained through:

1. The level of usage in the IS and the ICT awareness in the government- and business-level sector in Brunei
2. The level of awareness of green ICT in the government- and business-level sector in Brunei
3. The attitudes of government- and business-level sector in Brunei toward green ICT.
4. The awareness and implementation of BPR in the government- and business-level sector in Brunei for business sustainability.

The following chapter consists of a literature review, methodology, results, discussion, limitation, recommendation, and conclusion.

11.1.1 BUSINESS PROCESSING REENGINEERING (BPR)

BPR can also be known as business process change management, business process redesign, or business transformation, which is defined as a basic comprehensive rethinking or redesigning of all the organizational structure, job definitions, business processes, management systems, workflow, and the beliefs and assumptions of an organization by *The Business Dictionary.* According to the Hammer and Company's website, Michael Martin Hammer is recognized as one of the founders of management theory of BPR. The businesses and organizations in the market have been really competitive in increasing their profits in a world of business that is always changing and becoming more competitive. The main objective of a BPR is to change the previous way of performing task and working to a new one for a notable improvement or maximize profit through the use of ICT where it can help in the changing business methods. It became a key idea in the 1990s, where it influenced the management and transactions of organizations (Martin and Montagna, 2006). They further noted that it is not well accepted by organizations. In the first stage of the implementation, organization will be going through difficulties to get used to a new system as it is still new due to its complexity but their willingness to adapt to the new system will make a difference in the successfulness or the failure of the implementation. Stair and Reynold (2013) stated that the important aspect of an implementation is training where people need to be trained inthe application to avoid the mistakes that can be extremely costly to continue. They also suggested that procedures and policies have to go through the process of establishing, implementing, monitoring, and reviewing to reduce waste to maximize profits in a competitive world. Similar to procedures and policies, BPR needs to follow the process of establishing, implementing, monitoring, and reviewing to follow a successful implementation. Davenport and Short (1990) observed that companies that perform the major five steps of BPR were successful in the implementation. The implementation of five steps of BPR as suggested by Davenport and Short (1990) are as follows:

1. Develop the process objectives and business vision by setting targets and prioritize the objectives to improve the quality of work
2. Determine the processes to the redesigned by identifying the inefficient processes or critical processes
3. Measure and comprehend the existing problems by understanding and being able to identify the problems faced and setting a baseline. By identifying problems, companies can learn from their mistakes and prevent it from happening again while they come up with a baseline to adhere to the future.
4. Identify the information technology (IT) levers with the brainstorming of the new process approaches.
5. Outline and build a paradigm of the process by implementing the technical and organizational aspects.

However, Hall et al. (1994) identified four ways by which fail the BPR where they studied the damaging execution through the organizations that they have studied:

1. Allocating average performers. The redesigning process will fail when the performers fail to lead with credibility and skills.
2. Measuring the plan only. Companies that implement the system but do not follow nor track the system's process tend to fail.
3. Settling for the status quo. Organizations generally aim to expand the new redesign but they rarely translate the aspirations into a reality.
4. Overlooking the communication area. Organizations usually underestimate the power of communication in the implementation stage where it is essential to have a full communication program in order to succeed.

11.2 LITERATURE REVIEW

The widespread advancement and increasing growth of ICT havehelped economic development though as a consequence, the economic sustainability is affected. One of the main global concerns is to reduce the impact of the environmental footprint of the ICT industry as the use of computing devices plays a huge and unsustainable role in global warming and climate change, that the ICT industry is held accountable for an approximately 2%

of global carbon dioxide (CO_2) emissions. Thus, green ICT is the latest manifestation and a green alternative to achieving a more sustainable business practice where by definition "green ICT refers to an approach in reducing the energy and other resources consumed and the emissions and other waste produced across the ICT lifecycle—from manufacture, procurement, and use of ICT in an organization to its reuse and aims to improve environmental sustainability of organizations" (National Computer Board, 2011). To define green computing, Saha (2014) states that it is "the practice and procedures of using computing resources in an environment-friendly way while maintaining overall computing performance." It can also be considered as an ideal utilization and use ICT to oversee organizational activities in a more sustainable manner to achieve sustainable development.

Green IT is important to acknowledge the issues identified with the use of IT, mainly energy consumption, use of toxic substance, and e-waste (Hanne, 2011). First, the IT industry heavily relies on electricity for the operation of computers and datacenters, which cost operators such as Google and Amazon millions of dollars to run each year due to information becoming more digitalized and stored online (Reardon, 2009). Such companies that provide cloud computing which is described by Mell and Grance (2011) as "a model for enabling ubiquitous, convenient, on-demand network access to a shared pool of configurable computing resources that can be rapidly provisioned and released with minimal management effort or service provider interaction" is one of the fast-spreading technologies for users to store, manage, and process data though, however, it comes with a cost as in exchange for efficient and reliable services, data centers consumes a large amount of power (Sinha et al., 2011). Second, the manufacture and packing of hardware and software components and e-waste is a concern as toxic and hazardous substances such as lead, mercury, cadmium, and polychlorinated biphenyls are released when such consumer electronics break down (Velteet al., 2008), which presents a risk to the environment and can affect human health. The proper disposal of e-waste could be a challenge for developing countries where due to inadequate e-wastemanagement, they are buried, burnt in the open air, or dumped into surface bodies (Nnoromand Osibanjo, 2008). Each stage of a computer's life cycle from its production, use and ultimately, disposal, faces direct environmental challenges.

The adoption of green ICT is made by governments where environmental policies are aimed at reducing the environmental impact of ICT. Moreover, its practice offers organizations benefits and opportunities such

with 50,10, and 10%, respectively, while organizations A, E, and F chose none. Lastly, only organization A with 25% chose "never."

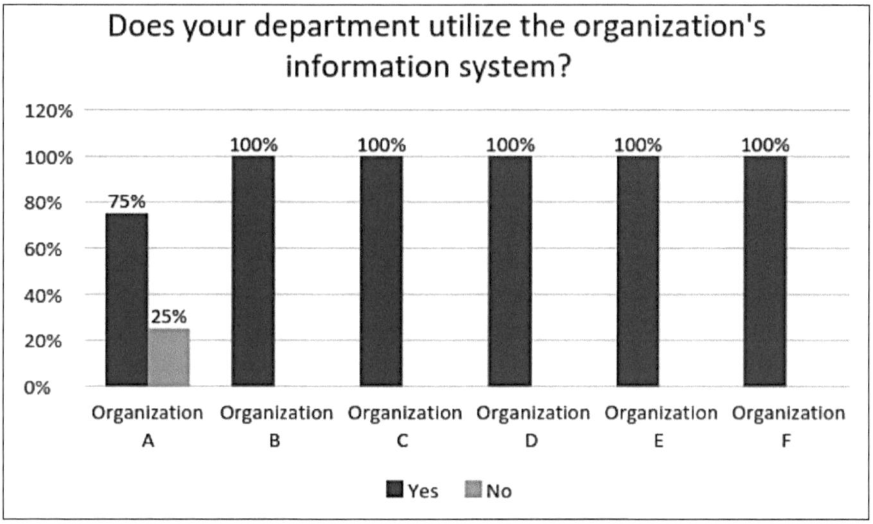

Figure 11.1 **(See color insert.)**

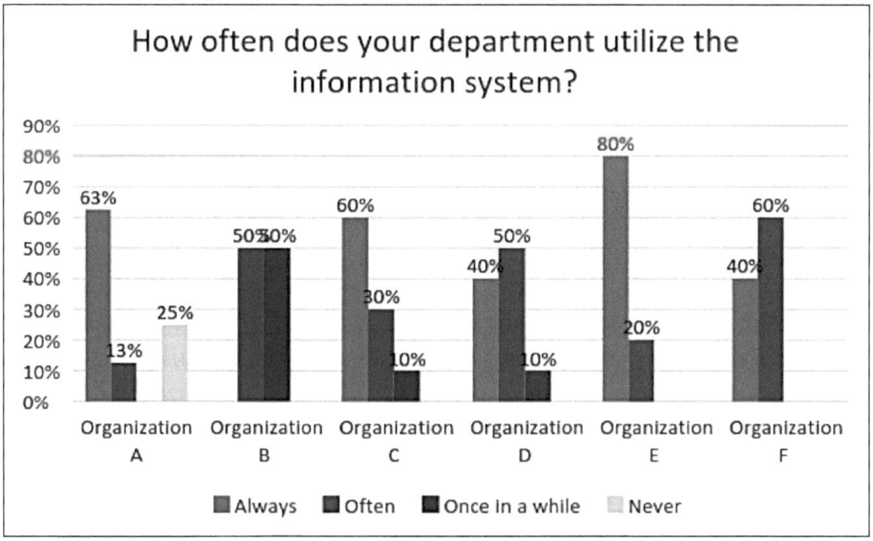

Figure 11.2 **(See color insert.)**

Figure 11.3 shows how often the management ISis reviewed by IT or the related people. All of the respondents chose "often (once every month)" for organization B and organization F. As for "Daily," only organizations A, C, D, and E chose it with 13,50, 30, and 40%, respectively. Organizations A, C, and D chose "others" with 50, 20, and 10%, respectively. None of the respondents chose "hourly."

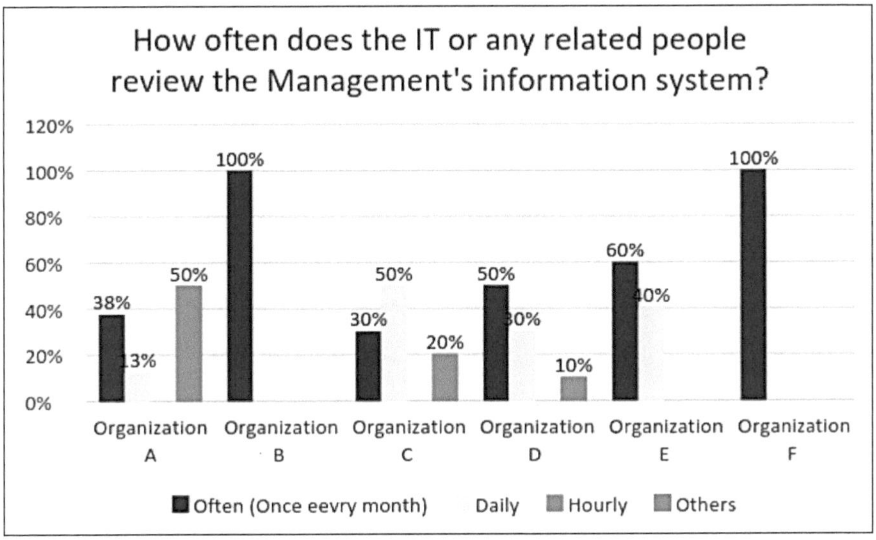

Figure 11.3 (See color insert.)

Figure 11.4 shows how long the organizations have been using the IS. According to the respondents from organization A, 13% chose "1–2 years" and 63% chose "2 years and more," while all respondents from organization B chose "2 years and more."From organization C, 10% of the respondents chose "6 months to 1 year," while 30% of the respondents chose "1–2 years,""2 years and more," and "less than 6 months." Organization D has an inconsistent response from respondents with 10% chose "Less than 6 months," 20% chose "6 months to 1 year," and "1–2 years." Lastly, organizations F and E havesimilar responsesfrom the respondents with 40% chose "1–2 years" and 60% chose "2 years and more."

Figure 11.5 shows if the respondents have the right authorization to access the documents in their IS. All the respondents have an access to the documents in their organization's IS for organization B. As for organization A, only 63% of the respondents chose "yes" and 38% chose "no." Organization C has a majority of 80% respondents choosing "yes," while the remaining percentchose "no." Organization D has an equal percentage of 50% respondents choosing "yes" and "no." As for organization E, majority of the respondents chose "no" and only 20% chose "yes." Lastly, organization F has 60% respondents choosing "yes" and only 40% chose "no."

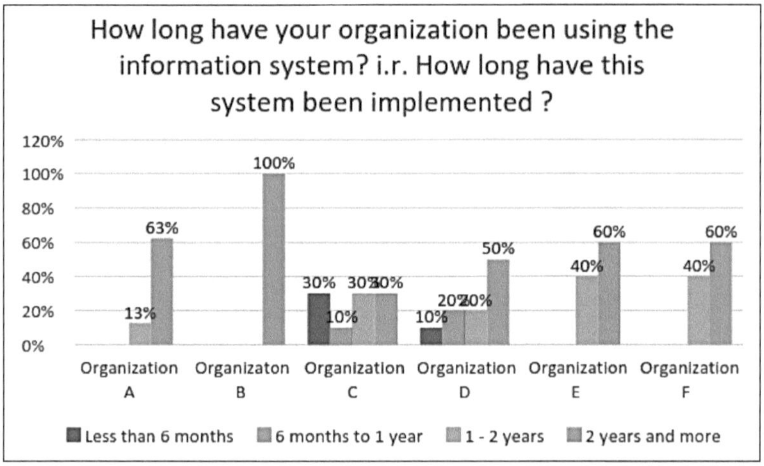

Figure 11.4 **(See color insert.)**

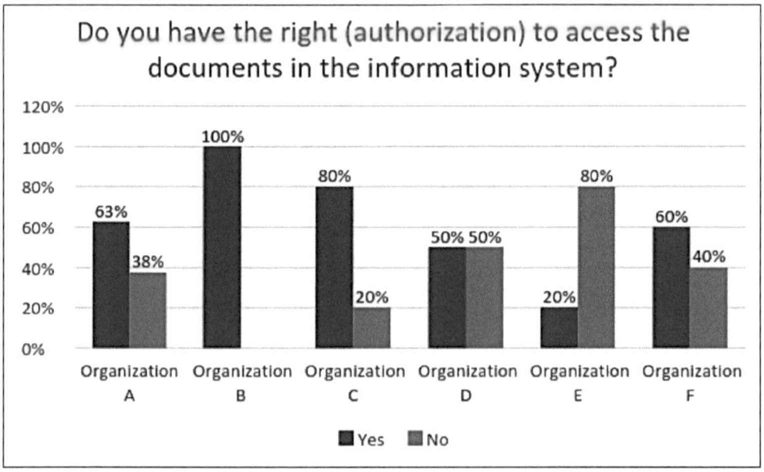

Figure 11.5 **(See color insert.)**

Figure 11.6 indicates what type of information security breach has their organization encountered. Option 1 is system failure/data corruption, option 2 is infection by viruses of malicious software, option 3 is theft or fraud involving computers, option 4 is attack by an unauthorized outsider/ system hacked, and option 5 is misuse of authorization access. Organization A has five respondents chose option 1, six respondents chose option 2, one respondent chose option 4, and two respondents chose option 5. Organization B has an equal number of respondents that chose options 1,2,4, and 5. Majority of organization C chose option 1 with eight respondents and five respondents chose option 3 and none for options 3,4, and 5. Organization D has an equal respondent of six for options 1 and 2, option 3 was chosen by only one respondent, option 4 was chosen by two respondents, and three respondents chose option 5. Organization E has respondents 4 out of 10 chose option 1, 5 chose option 2, none chose option 3, 2 chose option 4, and 5 chose option 5. Lastly, organization F has an equal number of respondents that chose option 1 and 4,that is, 3, 4 respondents chose the options 2 and 5, and none chose option 3.

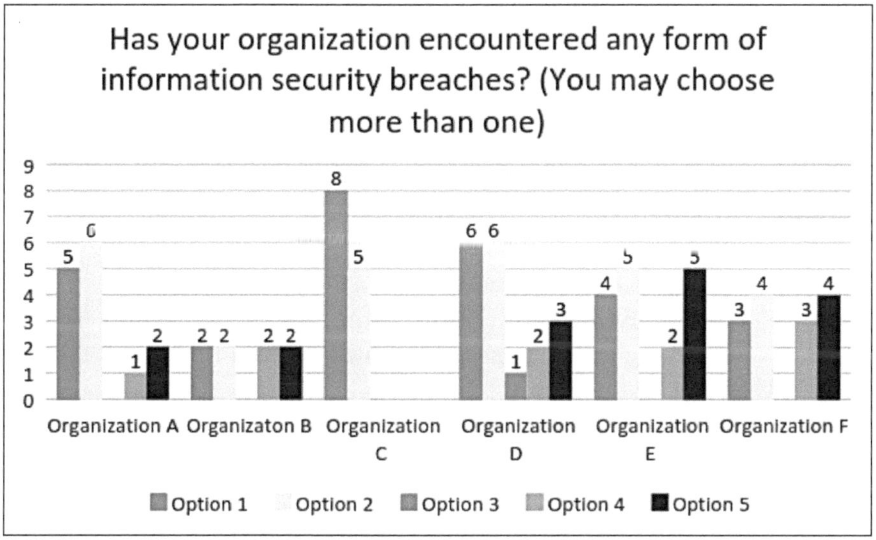

Figure 11.6 (See color insert.)

Figure 11.7 shows whether or not the organization outsources to external providers for the IS used irrespective of the organization they work in.Theanswers are not consistent from the respondents from the same organization except for organization B and organization F who chose "yes." For organization A, five out of eight respondents chose "yes," while the remaining chose "no." Organization C has an equal percentage,that is, 50% of the respondents choosing "yes," while the other 50% "no." Meanwhile, for organization D, only six out of nine respondents chose "yes." Lastly, for organization E, only two out of five respondents agree that their organization outsources for their IS.

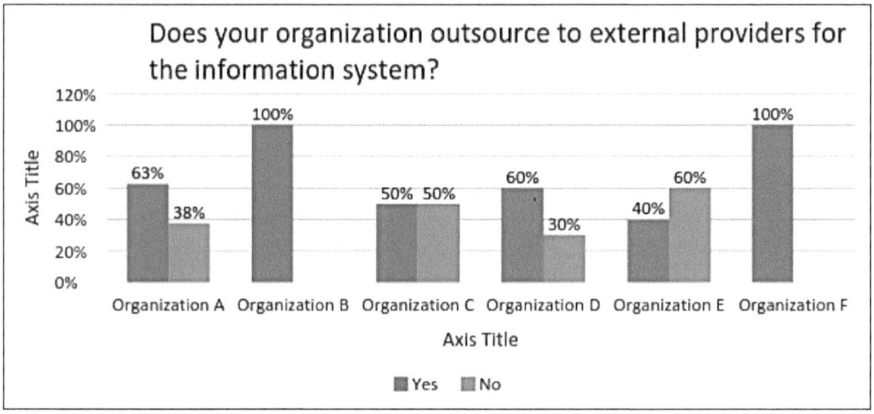

Figure 11.7 (See color insert.)

Most organizations turn to outsourcing to external providers for their information security system such as for its development, maintenance, and support in order to cut costs as it is cheaper to contract and hire third-party expertise. However, there is an inconsistency in response from the employees working in the same organization except for organizations B and F,while the rest have mixed results as to whether or not their organization does or does not. Based on the findings, this implies that the respondents from organizations A, C, D, and E are unsure or unaware of the organization's business practice.

In Figure 11.8, the graph compares the type of IS training that is provided in each respondent's respective organization. The options for the respondents to choose from are as follows:

Option 1: IS basics
Option 2: Management and implementation training in IS

Option 3: Application of information management/life cycle management
Option 4: Risk management and contingency planning
Option 5: None of the above

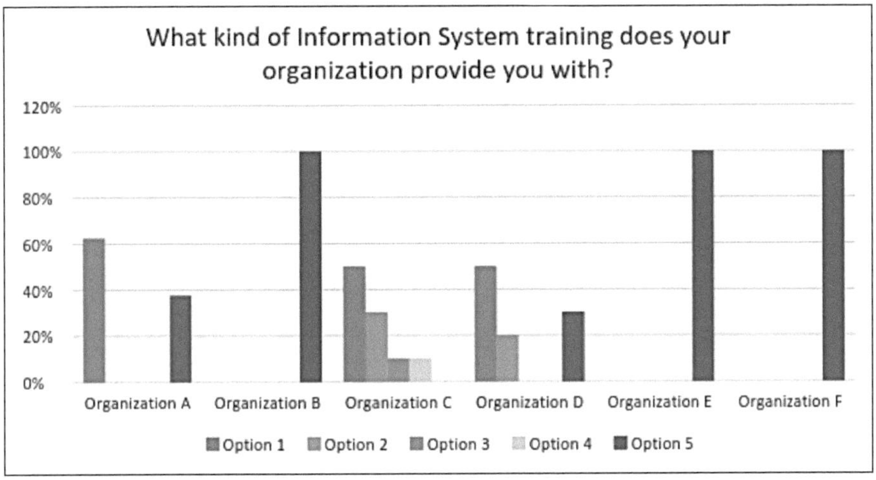

Figure 11.8 (See color insert.)

Figure 11.9 shows the results of the standard procedure in their respective organization or department present in their information security system. The respondents are allowed to choose more than one answer, where the options provided are as follows:

Option 1: Firewall
Option 2: Antivirus
Option 3: Formal written information security policy documented
Option 4: Implement authentication, authorization, accountingfor all users
Option 5: Accounts and passwords individually

For organization A, the most chosen options are options 1, 2, and 5,while the least chosen answers are options 3 and 4. For organization B, all of the respondents only chose options 4 and 5. However, for organization C, the respondents chose all but option 3. For organizations E and F, the respondents from both organizations only chose options 4 and 5.

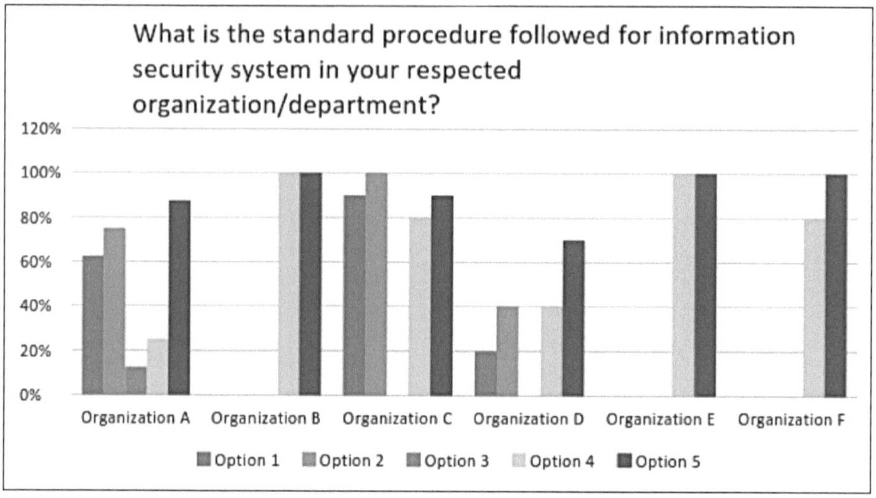

Figure 11.9 (See color insert.)

The standard procedures are essential for an IS for a more secure protection of information though, however, the types of standard procedures implemented vary differently as seen in the graph though there are different responses coming from the employees working in the same organization despite using the same IS (Susanto and Almunawar, 2018a; 2018b; 2015a; 2015b ; Liu et al., 2018 ; Susanto et al., 2011).

Figure 11.10 shows the results of the responses of each organization for the possible measures taken to mitigate information security system breach. The options are:

Option 1: Restrict access to the Internet at work
Option 2: Block access to inappropriate websites
Option 3: Block access to social networking sites
Option 4: Monitor websites/socials through logs

For organization A, the respondents havevaried answers as two out of seven chose option 1, three out of seven chose option 2, while the remaining two chose options 3 and 4, respectively. However, the respondents from organization B only chose option 1. Both organizations E and F havethe same results where they chose only option 2. As seen in the graph, the most picked options would be options 1 and 2. The least picked options are the remaining,that is, options 3 and 4,while for organization A, only 1 out of 10 respondents picked options 3 and 4, respectively. For organization B, only 2 out of 10 respondents picked option 4.

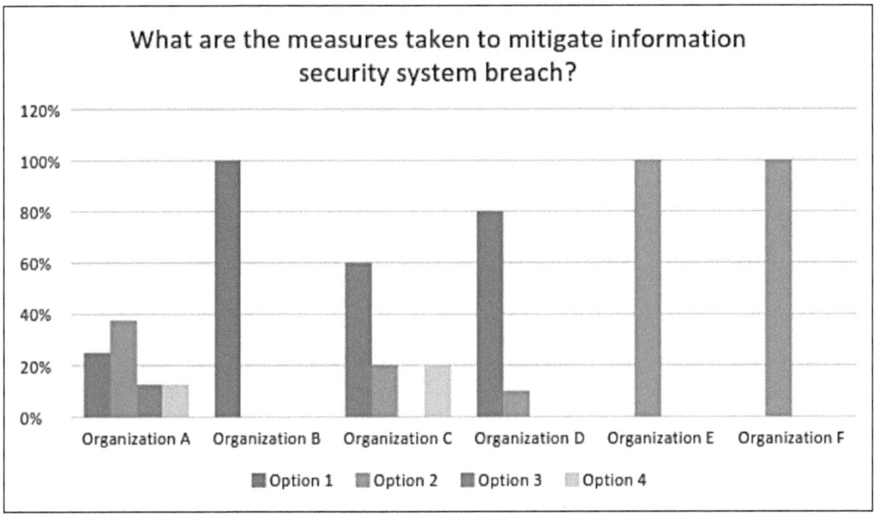

Figure 11.10 (See color insert.)

To ensure that the possibility of an information security system breach from happening, certain measures shouldbe taken such as the options listed above to avoid security breach or information leakage where the employees' activity would be restricted and monitored. The most picked option would be option 2, which blocks an access to inappropriate websites. However, as most or all of the options provided could be carried out or enforced in an organization, the question in the survey did not state that the respondents could choose more than one option. Thus, it may have limited the respondents from validly answering the question.

Figure 11.11 shows the options provided on how the organization could possibly react to the issues related to information security systems. They are as follows:

Option 1: Improvement in information security policies
Option 2: More investment in information security system
Option 3: Trainings on information security system

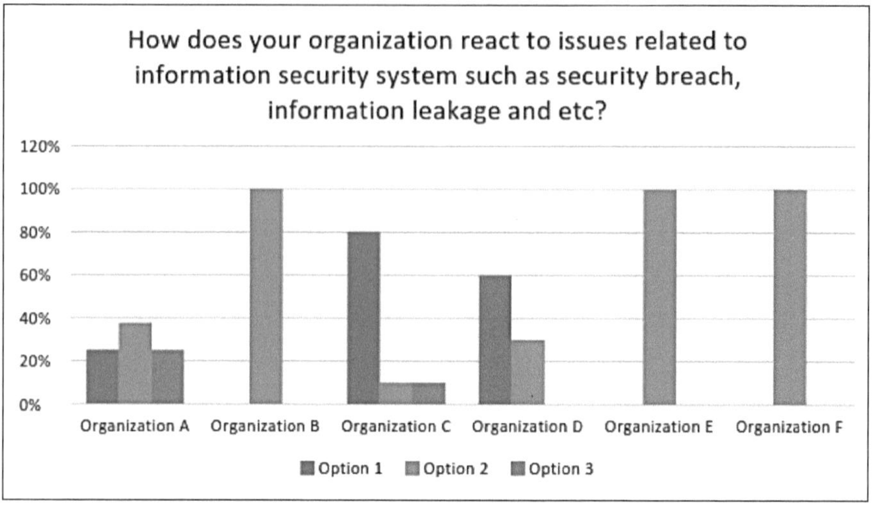

Figure 11.11 **(See color insert.)**

For organization A, seven out of eight respondents answered,while two out of seven chose option 1; three out of seven chose option 2, while the remaining two respondents chose option 3. For organization B, only option 2 was chosen. For organization C, however, 8 out 10 respondents chose option 1, while the remaining two respondents chose options 2 and 3, respectively. For organization D, six out of nine respondents chose option 1, while the remaining chose option 2. Organizations E and F havesimilar results where all the respondents from both organizations chose option 2 only.

The limitation on these types of questions in the survey is that it require respondents to only choose from the given options, that is, they are not able to provide their own answer if it was not given as an option and thus reducing the accuracy of the answers.

Figure 11.12 shows the results of the business processes that are involved by using the IS where the respondents are allowed to choose more than one answer and the options are as follows:

Option 1: Corporate website
Option 2: Corporate e-mail
Option 3: Payment processing
Option 4: Payroll processing

Option 5: Sales and/or marketing
Option 6: Customer transaction processing
Option 7: Finance and accounting
Option 8: Office tools (e.g., Word processing)
Option 9: Data storage

There is a similarity in the results for organizations B, E, and F, where all the respondents chose all the options except for options 1 and 2, possibly because the IS used in their organization provides all the information to do the provided business processes. There could be a possibility that their organization does not have their own corporate website or e-mail or the employees that participated in the survey are not aware. However, the answers vary with organizations A, C, and D. The most chosen option out of all would be option 9,that is data storage, which is vital as a part of the IS for the recording and storing of information such as crucial file systems, databases, and repositories. All of the mentioned business processes above are all equally important though it depends on the job position and requirements of the respondents who filled in the survey.

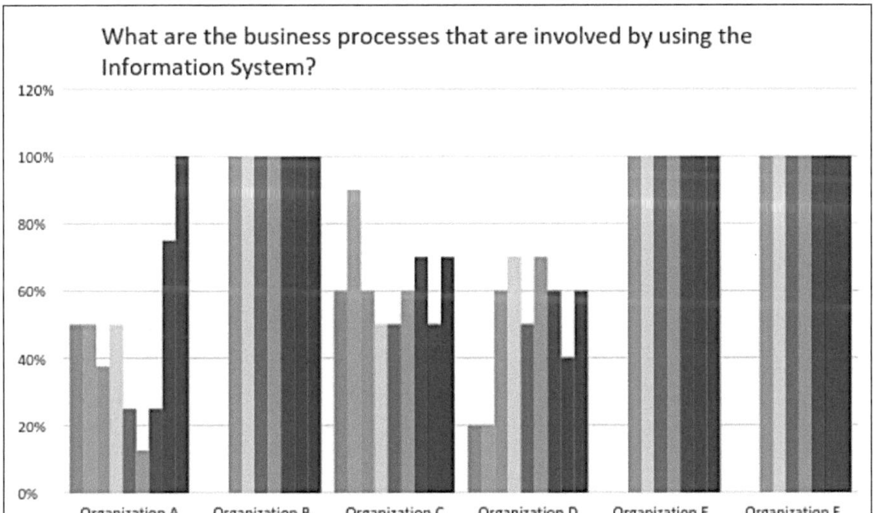

Figure 11.12 (See color insert.)

11.4.2 GREEN INFORMATION AND COMMUNICATION TECHNOLOGY (ICT) AWARENESS

The Figure 11.13 shows organizations A and B with equal respondents on the question set,"Are you aware of the risk to the environment because of global warming, carbon footprint, etc.?" with 50% chose either "yes" or "I am not aware." However, organization C was the most aware of the risk to the environment caused by global warming, etc. The other respondents among each organization had the highest awareness as well with 5 respondents among 10 in organization D and 3 respondents among 5 in organization E. Organization F had equalresponse for "yes" and "no, somewhat aware." All organizations had at least a minimum of one respondent who is not aware of what was asked in the question set in the questionnaire.

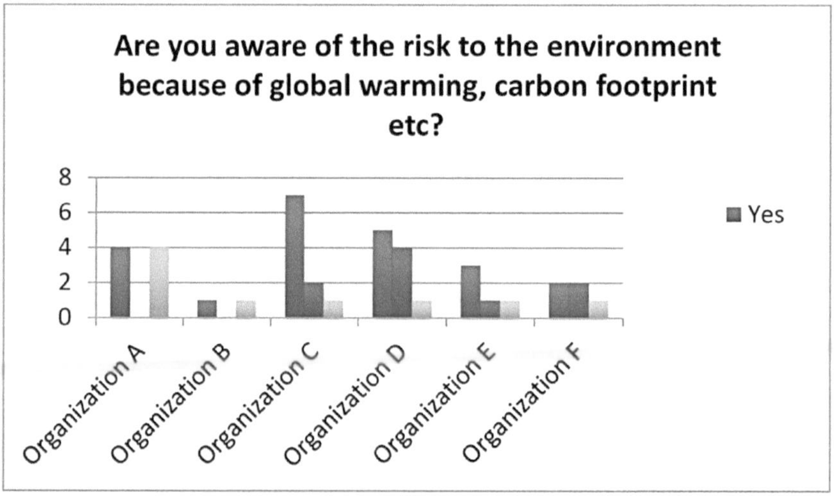

Figure 11.13 (See color insert.)

Figure 11.14 shows several ways in which green ICT deals with the use of computers and related technologies, which follows four different answers to choose from. They are:

Option 1: Minimizes energy and resource consumption
Option 2: Minimizes e-waste
Option 3: Reduces carbon footprint
Option 4: All of these

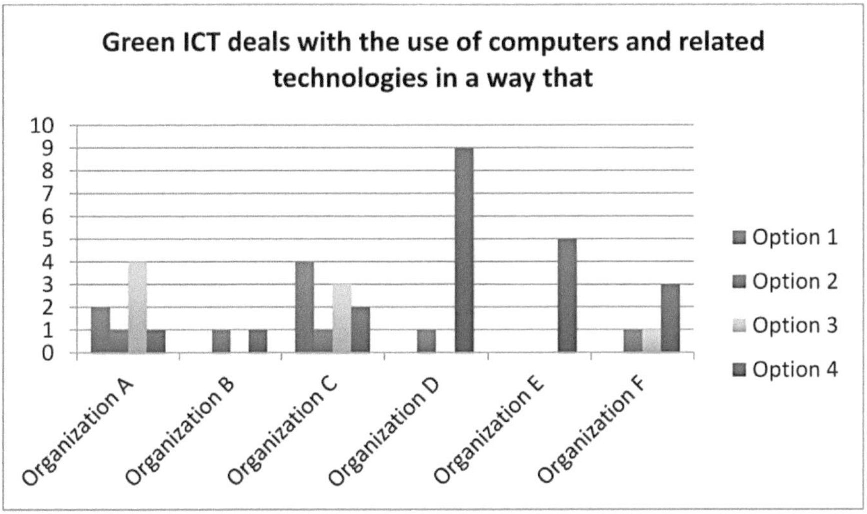

Figure 11.14 **(See color insert.)**

Organization A had the most respondents with option 3, followed by option 1 as 50 and 25%, respectively. The result follows with equal respondents with options2 and 4, respectively, of 13% each. Organization B chose equal parts of options 2 and 4. In organization C, four of the respondents chose option 1, followed by one respondent choosing option 2; three of the respondents chose option 3; andtwo of the respondents chose option 4. For organization D, 90% of the respondents went with option 4, while the remaining 10% went with option 2. However, for organization E, all respondents chose option 4. Lastly, organization F only chose options 2, 3, and 4 with equal percentage of 20% for options 2 and 3, and the remaining 60% went with Option 4.

Figure 11.15 shows the average laptop uses 20% as much energy as a desktop PC" in the questionnaire. This question gives options as true or false. For organizations A, B, and C show equal percentage of 50% of both, true and false, where four respondents each from organization A, one respondent each for true and false for organization B, and five respondents each for organization C chose an equal part of either true or false. Organization D on another hand had 60% for false and the remaining for true. Organization E was opposite and had a higher percentage of 80% for true and the remaining 20% chose false. For organization F, two respondents chose true, whereas the remaining three out of five respondents chose false.

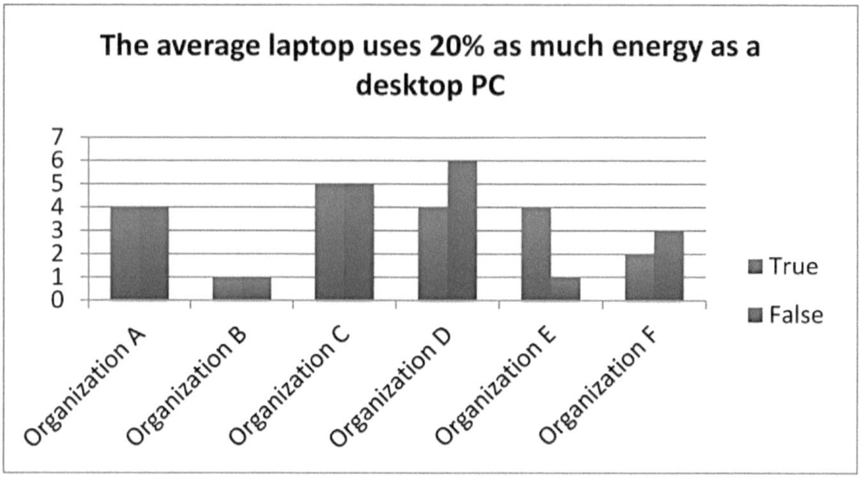

Figure 11.15 (See color insert.)

In green ICT awareness, the respondents were asked the questions based on their knowledge about computer's processors where the question was phrased as "As personal computer's processor gets more powerful, they likewise increase their power demands." Participants were required to indicate true or false as their answers on the provided questionnaire. The results were such that in organization A, seven respondents with 88% answered true and one participant with 13% answered false. In organization B with two participants only, they each answered true and false, leaving the results to an equal of 50% false and 50% true. In organization C, there were 10 participants, and 90% of the 10 participants answered true, while the other 10% answered false. In organization D, there were 10 participants, and seven respondents with 70% answered yes, while the other three respondents with 30% answered false. Furthermore, in organization E, there were only five participants and 100% answered true. In addition, organization F also had five participants, where three out of five with 60% answered true and the other two with 40% answered false.

Participants were further asked abouttheir awareness of power consumption inthe situation wherethey switch on and off or restarting multiple times, the power consumed increases rather thanleaving it switched on. The question in the predetermined questionnaire was "Constantly shutting down and restarting your computer during the day would consume more energy than leaving it running." Participants were also required to indicate true or false on the survey distributed to them. The results from six organizations were as followed: In organization A with eight participants, four, that is, 50% answered "yes," while the other four respondents answered false, causing it to be divided equally in 50%. However, organization B had only two participants and one answered true and the other one answered false, causing it to be equally divided similar to organization A. Furthermore, organization C had ten participants, where eight answered true, leaving it a percentage of 80% and the other 20% was two respondents that disagreed. Similarly, organization D also had ten participants, where seven respondents answered true with a percentage of 70%, while the false was answered by three respondents with 30%. Furthermore, organization E had five participants, where three answered true, with a percentage of 60% and the other two respondents with a percentage of 40% with the answer false. Lastly, organization F's response on this question was 60% true with three out of five participants and 40% false with the other two participants.

In another question from the questionnaire, "It is better to use your equipment as long as possible to keep it from becoming e-waste." The participants were asked abouttheir awareness on e-waste where the statement suggested that it is more environmental friendly to recycle and keep using equipment rather than throwing it and let it become an e-waste. The results were that in organization A with eight participants, four respondents answered true, setting the percentage rate to be 50%, while the other four respondents answered false, also setting a percentage rate of 50%. Further, in organization B with only two participants has a result of 50% true and 50% false with one each answering true and false, respectively. Additionally, in organization C, four participants out of ten agreed that it is better to continue using the equipment causing a percentage of 40%, while the other 60% disagreed with the statement. Meanwhile, in organization D, five out of ten participants answered true with 50%, 40% with four participants answered false, while one participant equivalent to 10% did not answer the question in the questionnaire. Organization E had five respondents, where two answered true, while two answered false, leaving a percentage of 40% and 60%, respectively. Lastly, organization F also had five participants, but the results are different as three respondents indicated true, while the other two indicated false, setting a percentage of 60 and 40%, respectively.

The below graph (Figure 11.16) shows organization A with a total of eight respondents, where seven chose true and the remaining respondent chose false. For organization B, all respondent chose true. For both organizations, C and D, with a total of 10 respondents each, both had 7 respondents choosing true and 3 respondents choosingfalse. The total does not add up for both Organizations as 1 from each organization left this particular question blank. However, for organizations E and F, four out of five respondents chose true and three out of five respondents chose true, respectively. The remaining chose false.

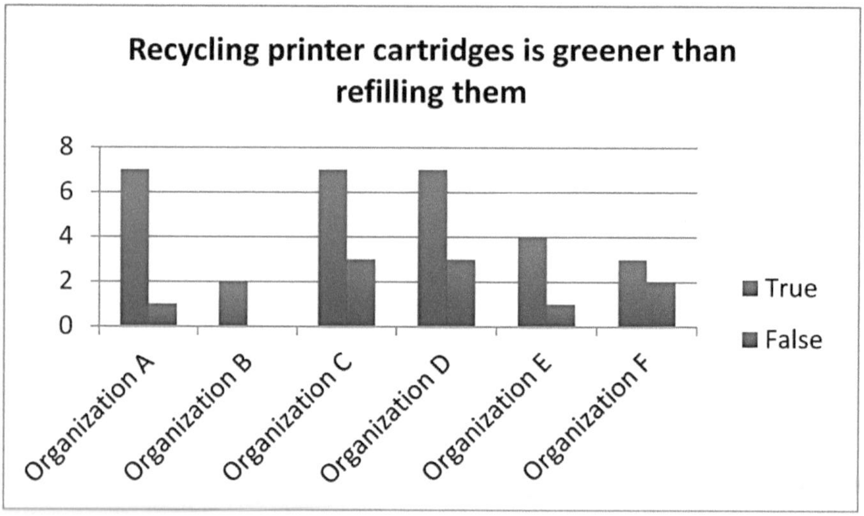

Figure 11.16 (See color insert.)

The questionnaire further asked participants' knowledge on a way to reduce carbon footprint as they stated that using online learning or educational services could reduce the carbon footprint. The question printed on the questionnaire was "Online deployment of learning and educational services is a way to reduce carbon footprint." In organization A, six out of eight respondents answered true with 75% and 25% with two respondents answering false. Furthermore, organization B's two participants answered false, setting a percentage of 100%. For the other organization, organization C, ithad a percentage rate of 80%,where eight answered true and 10%,where one answeredfalse, while one respondent did not indicate true or false in the questionnaire. Additionally, organization D responded with

60% true with six participants and 40% false with four participants. Inorganization E, two out of five participants agreed with the statement, setting a 40% percentage rate, while three out of five participants disagreed with the statement with a percentage rateof 60%. In organization F, three out of five participants indicated true and two out of five indicated false, with a percentage rate of 60 and 40%, respectively.

Participants were also asked whether they think that telecommuting is a mean of working or learning which can reduce the traveling. The question predetermined on the questionnaires handed out was, "Telecommuting means to work or learn from home and thus reduce travel." The results that were obtained from organizations A, B, C, D, E, and F are as followed: In organization A, all eight respondents agreed with the statement by indicating true, with a percentage rate of 100%. In organization B, there were two respondents and both the respondents also indicated true in the questionnaires, with a percentage rate of 100%. Furthermore, in organization C, seven out of ten participants agreed with the statement by indicating true, setting a percentage rate of 70%, while two participants set a percentage rate of 20% by indicating false but one participant did not answer the question by leaving it blank, which contributed to a percentage of 10%. Meanwhile, in Organization D, seven out of ten respondents indicated true, while three out of ten answered false, setting a percentage rate of 70 and 30%, respectively. In organizations E and F, which have five participants each, all the participants from both the organizations indicated true and agreed to the statement, setting a percentage rate of 100% of both the organizations.

The chart of Figure 11.17 shows that for organizations B and F two and five respondents, respectively, chose all false for this question set by in the questionnaire. Organization A has a low 13% for true and a high 88% for false. For organizations C and D, both organizations had similar responses, where seven of each organization's respondents chose true and the remaining three respondents from each organization chose false. Lastly, for organization E, 40% of the respondents chose true and the remaining 60% chose false.

The last question for "Green ICT Awareness" asked was if "Studying electronic documents is greener than printed documents." Organizations B, E, and F, all of which have two, five, and five respondents, respectively, each chose true. For organization A, 88% chose true and the remaining percentage chose false. In organization C,out of 10 respondents, 8 respondents chose true and 1 respondent chose false. One respondent from organization C did not respond to this question. Lastly, in organization D of 10 respondents, 90% chose true and the remaining 10% chose false.

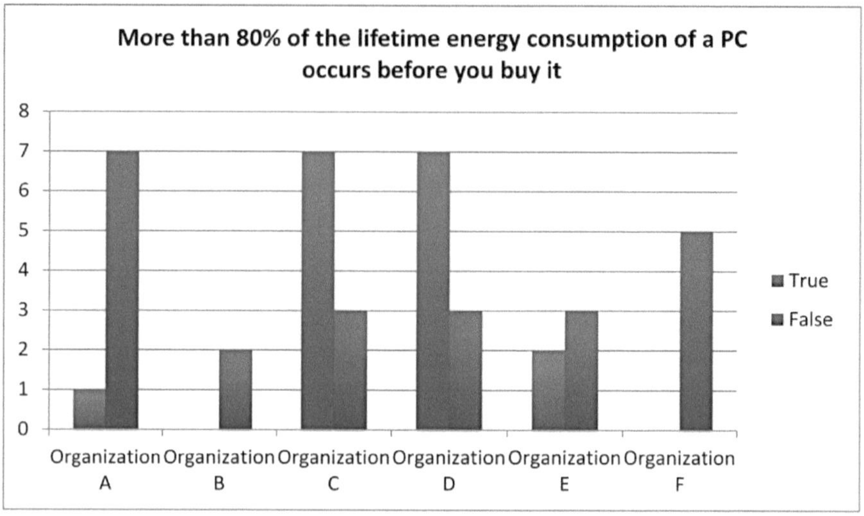

Figure 11.17 **(See color insert.)**

11.5 DISCUSSION

11.5.1 GREEN ICT AWARENESS

In Green ICT Awareness, it can be concluded that the respondents collectively had mixed views regarding Green ICT Awareness. According to the results, it shows the proof that green ICT is not common. In Figure 11.13, organization C shows the most awareness although not being a strong IT field. However, in this research, the public organizations show equal awareness. With 50% each shown in Figure 11.13, it can be concluded that from the public sector and from the respondents that green ICT has been proven to spread but falls short to the knowledge for everyone.

Moving on to the results shown in Figure 11.14, one of the options given was minimizing energy and resource consumption, which, from the sentence proves to save more energy in terms of power and other forms of energy. The other included the minimizing of e-waste from green ICT that could harm the environment. Option 3 reduces carbon footprint, which in other terms means that it reduces the amount of greenhouse gases. Although organizations A and B, which are from the public sector show a little proof to the usage of green ICT, organization A agrees more on the reduction of carbon footprint by using green ICT. However, the results shown in Figure 11.14 shows that organizations D, E, and F, which are from the private sectors claim to agree that all of the options given deal with the use of computers and other related technologies that helps with the awareness of green ICT.

The average laptop is said to have consumed 20% more energy than a desktop personal computer. Generally, this fact could be both true and false. A personal computer functions with running power. This means it can only function through a plugged socket, whereas a laptop consumes energy from only the portability of the device itself. A laptop is able to function without consuming continuous power electricity. Besides the fact that the power used to charge a laptop comes from generating electricity. This power can possibly use up more energy to charge the device than using a desktop personal computer. However, the same energy is used when a desktop personal computer is used longer than the usual period. From the Figure 11.15, 50% of the respondents agree and disagree to this fact.

Recycling printer cartridges is greener than refilling them. Green ICT from the title itself shows that saving the environment is better than any other option. For one, recycling the cartridges helps in the reduction of wastes. This can be seen in Figure 11.16 that all organizations agree with recycling the cartridges is greener than refilling them as the recycled products are involvedingood use. All other questions set in the questionnaire are facts to show if organizations are aware of green ICT in the country. Many of the factsshow both true and false according to the respondents themselves as this section was to see if people were aware of green ICT.

11.5.2 ATTITUDES TO GREEN ICT

The uptake of IT leads to the even more generation of e-waste. It is right for people to know the significant risk of consuming energy and generating e-waste. In Figure 4.1, the question was asked whether the respondents think that IT community has contributed to put the environment at such risk. From the result shown in the graph, the respondents' answers are mostly skewed to the left side, which means that majority agree that IT community has been involved in putting the environment at risk. Organizations C, D, and E are with the highest agreements, where 80% of the respondents are from organizations C and D and 100% from organization E. Both public and private organizations agree that IT is the main source of computer usage, generating e-waste because business sectors are most likely to have ICT items such as PC, monitor, and laptop. The composition of the components of these items is considered to be hazardous, presenting elements such as mercury, arsenic, and lead, which can contribute to the harming of environment.

The result shown in Figure 4.2 shows that the organizations believe that the large IT companies are doing enough to integrate policies and solutions that are environment-friendly. For example, as observed in Figure 4.3, 70% from organization D agreed that the large IT companies have done enough. However, some organizations such as organization A has a slight disagreement of 12.5%, and 50% that has no idea whether large IT companies are doing enough or not. This result is maybe because large IT companies have not done enough for everyone, including the respondents to realize that they have done anything to integrate environmental policies and solutions. This lack of existing policies may affect the attitude toward the implementation of green ICT.

Due to e-waste generation and greenhouse gases being produced with 73%, which is more than half of the total respondents to agree that the IT community has contributed to put the environment at risk (Figure 4.1). This leads to 83% of the respondents from all six organizations to strongly agree that educational institutes should take the initiative in spreading green computing awareness and implement it as a practice (Figure 4.3). It is important for IT personnel to update the knowledge with regard to green ICT because the lack of skills may hinder the implementation of green ICT to be successful. Educational institutes should support this implementation by means of having seminars or workshop regarding it to encourage other people, not just IT personnel to know about green ICT.

The increase of carbon footprint increases the production of greenhouse gases is significantly risking the environment. Figure 4.4 shows whether carbon footprint can be reduced if using learning methods and services when implementing ICT. It is observed that 75% agreed that it helps institutes to reduce carbon footprint. Apart from helping people to be aware of green ICT, the institutes can put on some action by implementing services and methods so that carbon footprint can be helpedto be reduced. Methods such as recycling the ICT items or using these items efficiently, making the most out of it without any wastage, can be used.

In all sectors of the economy ICT was introduced, and this has caused the usage of ICT to increase. This increase may affect the climate in the business chain system because ICT consumes large amount of energy and emits greenhouse gases, which are very hazardous. Green ICT has been used to sustain IT in every environmentally way possible. The primary target is to reduce the large consumption of energy and the hazardous greenhouse gases that ICT has caused. There are a few recommendations that can potentially be a way to improve the environmental efficiency of ICT.

As from the surveys conducted, the results overall showed that ICT personnel are not aware of the impacts that green ICT makes and what the advantages it can gain to organizations. ICT personnel play an important role here because helping them to be aware of green ICT is going to likely be the drivers of green ICT uptake. If ICT personnel havethe knowledge to use it, they should be prepared for its implementation. It is very important for people to understand that everyone should know and be aware of green ICT, particularly, business organizations so that they can engage with it and know potentially the chance of getting a sustainable world that all aimed for. However, if there is any disparity in the understanding of implementing green ICT, it will be difficult.

From the result in Figure 11.13, it is observed that many responses were strongly skewed to the agreement that they are aware of green ICT. However, they are not aware if their organization has implemented green ICT or not. This shows that it is possible that there is a lack of training in green ICT or maybethere has never been any training at all. This lack of training can lead to the barriers toward green ICT to be implemented. This also provides the reason why, according to our findings, the organizations have no idea whether they have been using green ICT or not. This lack of training, too, results in the lack of skills where IT staff are not able to support the implementation of green ICT or let alone any innovative implementations. Training is vital for any organizations because if the skills are lacking, workers' productivity will be limited and therefore limiting the application of any new technology. Organizations may result in having to face challenges from it. Therefore, planned training should be done and all personnel of the organizations who are involved in green ICT implementation are required to participate.

Another way to let people be aware of green ICT is to conduct workshops and seminars about it so that ICT personnel are updated about the green ICT knowledge. Thereason why organizations are not aware of green ICT is that because there are no workshops or seminars to let them know about it. By means of creating awareness, workshops and seminars may help personnel to improve the skills of ICT and their expertise levels, therefore increasing employees' productivity. The third way is to manage the organizations, where implementing new innovations like green ICT will result in them being resistant to the change becauseany implementation of new technologies depends on the people. People resist the changes of their existing work practices or due to high cost and mainly due to the lack of awareness and the perceived feeling of not needing the skills to implement and support such technologies. Therefore, such resistance has to be managed by the top-level management to make sure that green ICT implementation will be successful. The fear of failure to realize what e-waste causes from the uptake of ICT can be another reason forthe lack of awareness. As from Figure 11.13, it is observed that few respondents from all six organizations had no idea what IT has contributed to. It is important for everyone to be aware because the lack of support for the implementation of green ICT can hinder the process of it to be successful.

KEYWORDS

- **business**
- **environment**
- **e-waste**
- **greenhouse gas**
- **recycling**

REFERENCES

Ahmed, F.; Naeem, M.; Iqbal, M. ICT and Renewable Energy: a Way Forward to the Next Generation Telecom Base Stations. *Telecommun. Syst.* **2016,** *64*(1), 1–14.

Ahola, J.; Ahlqvist, T.; Ermes, M.; Myllyoja, J.; Savola, J. ICT for Environmental Sustainability: Green ICT Roadmap; VTT Tiedotteita, Finland–Research Notes, 2532, 2010.

Almunawar, M. N.;Anshari, M.;Susanto, H.; Chen, C. K. Revealing Customer Behavior on Smartphones.*Int.J. Asian Business Information Management* **2015a,** *6*(2), 33–49.

Almunawar, M. N.; Susanto, H.; Anshari, M.The Impact of Open Source Software on Smartphones Industry.In *Encyclopedia of Information Science and Technology,* 3rd ed.; Khosrow-Pour, M., Ed.; IGI Global,2015b, pp 5767–5776.

Almunawar, M. N.; Anshari, M.; Susanto, H.; Chen, C. K. How People Choose and Use Their Smartphones. In *Management Strategies and Technology Fluidity in the Asian Business Sector;* Ordóñez de Pablos, P, Ed.; IGI Global, 2018a, pp 235–252.

Almunawar, M. N.; Anshari, M.; Susanto, H. Adopting Open Source Software in Smartphone Manufacturers' Open Innovation Strategy.In *Encyclopedia of Information Science and Technology, 4th ed.;*Khosrow-Pour, M., Ed.; IGI Global, 2018b, pp 7369–7381.

Awan, U.;Sarwar, S.;Raza, M. A. Green Consumer Behavior and Environmental Sustainability. *J. Sci. Technol.* **2011,** *1*(1), 1–3.

Davenport, T. H.; Short, J. E. The New Industrial Engineering: Information Technology and Business Process Redesign. *Sloan Manag. Rev.* **1990,** *31*(4), 11–27. http://search.proquest.com/docview/224963315?accountid=9765</bib>

Griffith, E.How to Recycle your Technology. The PC Mag. http://www.pcmag.com/article2/0,2817,2383568,00.asp (accessed January 22, 2016)

Hall, E. A.; Rosenthal, J.; Wade, J.How to Make Reengineering Really Work. *McKinsey Quarterly* 2, 1994, pp 107–129.

Hanne, F. Z. Green-IT: Why Developing Countries Should Care. *Int. J. Comp. Sci. Issue* **2011,** *8,* 147–153.

Leu, F. Y.; Liu, C. Y.; Liu, J. C.; Jiang, F. C.; Susanto, H. S-PMIPv6: An Intra-LMA Model for IPv6 Mobility. *J. Network Computer Applications* **2015,** *58,* 180–191.

Leu, F. Y.; Ko, C. Y.; Lin, Y. C.; Susanto, H.; Yu, H. C. Fall Detection and Motion Classification by Using Decision Tree on Mobile Phone. In *Smart Sensors Networks;*Xhafa, F., Leu, F.-Y., Hung, L.-L., Eds., Academic Press,2017, pp 205–237.

Liu, J. C.; Leu, F. Y.; Lin, G. L.; Susanto, H.An MFCC-Based Text-Independent Speaker Identification System for Access Control.*Concurrency Computation: Practice Experience* **2018,** *30*(2), e4255.

Martin, R. L.; Montagna, J. M. Business Process Reengineering Role in Electronic Government. In The Past and Future of Information Systems: 1976–2006 and Beyond; Avison, D.; Elliot, S.; Krogstie, J.; Pries-Heje, J., Eds.; Springer: U.S., 2006, pp 77–88).

Mell, P.; Grance, T. The NIST Definition of Cloud Computing. *Commun. ACM* **2011,** *53*(6), 50.

Murugesan, S. Harnessing Green IT: Principles and Practices. *IT Prof.* **2008,** *10*(1), 24–33.

National Computer Board.*Green ICT Guidelines for Businesses*. National Computer Board: Mauritius, 2011.

Nnorom, I. C.; Osibanjo, O. Overview of Electronic Waste (e-Waste) Management Practices and Legislations, and Their Poor Applications in the Developing Countries. *Res. Conserv. Recycl.* **2008,** *52*(6), 843–858.

Reardon, M. How Secure is the US Communications Network? *CNET News*. **2009**. http://www.newageitsolutions.com/pdf%20files/April%202009/Week%203%20updates/How%20secure%20is%20the%20U.S.%20communications%20network.pdf</bib>

Saha, B. Green Computing. *Int. J. Comp. Trends Technol.* **2014**, *14*(2), 46–50.

Sinha, R.; Purohit, N.; Diwanji, H. Energy Efficient Dynamic Integration of Thresholds for Migration at Cloud Data Centers. *Int. J. Comp. Appl.* **2011**, *1,* 44–49.

Stair, R.; Reynolds, G. Principles of Information Systems: the Personal and Social Impact of Computers. Cengage Learning: U.S. **2013**.

Susanto, H. Managing the Role of IT and IS for Suppoting Business Process Reengineering. *J. Systems Information Technol.* **2016a.**

Susanto, H. IT Emerging Technology to Support Organizational Reengineering. **2016b**. https://ssrn.com/abstract=2770318

Susanto, H. Cheminformatics—The Promising Future: Managing Change Of Approach Through ICT Emerging Technology. In *Applied Chemistry and Chemical Engineering, Volume 2: Principles, Methodology, and Evaluation Methods*; Haghi, A. K., Pogliani, L., Balkose, D., Mukbaniani, O. V., Mercader, A. G., Eds.; Apple Academic Press, 2017a, pp 313–332.

Susanto, H. Biochemistry Apps as Enabler of Compound and DNA Computational: Next-Generation Computing Technology. In *Applied Chemistry and Chemical Engineering, Volume 4: Experimental Techniques and Methodical Developments;*Haghi, A. K., Pogliani, L., Castro, E. A., Balköse, D., Mukbaniani, O. V., Chia C. H., Eds.; 2017b, Apple Academic Press, pp 181–202.

Susanto, H. Electronic Health System: Sensors Emerging and Intelligent Technology Approach. In *Smart Sensors Networks;*Xhafa, F., Leu, F.-Y., Hung, L.-L., Eds., Academic Press, 2017c, pp. 189–203.

Susanto, H. Smart Mobile Device Emerging Technologies: an Enabler to Health Monitoring System. In *High-Performance Materials and Engineered Chemistry;* Torrens, F., Balköse, D., Thomas, S., Eds.; Apple Academic Press, 2018, pp 241–264.

Susanto, H.; Almunawar , M. N. Managing Compliance with an Information Security Management Standard. In *Encyclopedia of Information Science and Technology,* 3rd ed.; Khosrow-Pour, M., Ed.; IGI Global, 2015, pp 1452–1463.

Susanto, H.; Almunawar, M. N. *Information Security Management Systems: A Novel Framework and Software as a Tool for Compliance with Information Security Standard;* CRC Press, 2018.

Susanto, H.; Chen, C. K. Macromolecules Visualization through Bioinformatics: an Emerging Tool of Informatics. *Appl. Phys. Chem. Multidiscip. Approaches* **2018a,** 383.

Susanto, H.; Chen, C. K. Informatics Approach and Its Impact for Bioscience: Making Sense of Innovation. *Appl. Phys. Chem. Multidiscip. Approaches* **2018b,** 407.

Susanto, H.; Almunawar, M. N.; Tuan, Y. C. Information Security Management System Standards: a Comparative Study of the Big Five. *Int. J. Electrical Computer Sci.***2011,** *11*(5), 23–29.

Susanto, H.; Chen, C. K.; Almunawar, M. N. Revealing Big Data Emerging Technology as Enabler of LMS Technologies Transferability. In *Internet of Things and Big Data Analytics Toward Next-Generation Intelligence;*Dey, N., Hassanien, A. E., Bhatt, C., Ashour, A. S., Satapathy, S. C., Eds.; Cham: Springer, 2018, pp 123–145.

Velte, T.; Velte, A.; Elsenpeter, R. C. Green IT: Reduce your Information System's Environmental Impact While Adding to the Bottom Line. McGraw-Hill, Inc.: New York, NY, 2008.

INDEX

D

E

F

W

Y

Z